H Holt-Butterfill

First Principles of Mechanical and Engineering Drawing

H Holt-Butterfill

First Principles of Mechanical and Engineering Drawing

ISBN/EAN: 9783744749978

Printed in Europe, USA, Canada, Australia, Japan

Cover: Foto ©berggeist007 / pixelio.de

More available books at **www.hansebooks.com**

FIRST PRINCIPLES

OF

MECHANICAL

AND

ENGINEERING DRAWING

A COURSE OF STUDY ADAPTED TO THE SELF-INSTRUCTION OF
STUDENTS AND APPRENTICES TO MECHANICAL
ENGINEERING IN ALL ITS BRANCHES

AND FOR THE USE OF TEACHERS IN TECHNICAL AND MANUAL
INSTRUCTION SCHOOLS

BY

H. HOLT-BUTTERFILL, M.E.

FORMERLY A MEMBER OF THE INSTITUTION OF MECHANICAL ENGINEERS AND
INSTITUTION OF NAVAL ARCHITECTS

*WITH UPWARDS OF 350 DIAGRAMS IN ILLUSTRATION OF THE
PRINCIPLES OF THE SUBJECT*

LONDON: CHAPMAN AND HALL, LIMITED

1897

PREFACE

THE greater part of the subject matter of this book appeared in a series of articles in the *Mechanical World*. The purpose in writing it is so fully explained in the Introduction that a Preface is hardly required. As the forms given to the various parts of a machine or engine are on analysis invariably found to be combinations of certain geometrical solids, a knowledge of how each of these should be drawn when in any position should be first acquired by the student draughtsman. To this end a series of problems is given in the following pages, commencing with the construction of those simple geometrical figures which form the surfaces of the solids which give shape to mechanical details, and subsequently the method adopted in representing the solids themselves, singly and in combination.

As no amount of *copying* "drawings" of mechanical details will ever give the student a knowledge of the reasons why they are made to take the special forms given to them, so in the earlier stages of the study of mechanical drawing it is impossible for him to acquire the power to draw the simplest solids in different positions correctly without a knowledge of the principles of "Orthographic Projection," which is the basis of the representation of all solid objects. In this part of the subject an extended series of problems is given, the solution of which should enable the student to draw any simple object without further help.

In the method of studying the contents of this work, the student is advised to take the different parts of the subject in the order in which they are arranged, as he will thereby be led to acquire a mastery of it in a way that will impress upon his mind the connection that each part bears to that which follows. The order of study may not be that usually followed, but it is such as an association of many years with draughtsmen and students has proved to the author to be the best for the acquisition of the preliminary knowledge necessary to the successful practice of the draughtsman's art.

This work is not intended as a treatise on either Plane or Solid Geometry, but as much of these subjects is given as will be required by the student to attain to an easy comprehension of the first principles of mechanical drawing as herein exemplified. Their actual application to the delineation of machine elements and engine details may possibly form the subject of a further work.

H. HOLT-BUTTERFILL.

Greenwich, 1897.

CONTENTS

Introduction.—THE VALUE OF A KNOWLEDGE OF DRAWING TO THE STUDENT

PAGE

CHAPTER I

THE TOOLS AND MATERIALS REQUIRED BY THE STUDENT

Drawing-Board—Tee-Square—Adjustable Bladed Square—Set-Squares—
Pencils—Drawing - Pins—Paper—Rubber—Ink—Drawing Instruments   ...   .   1—11

CHAPTER II

MECHANICAL AND FREEHAND DRAWING : THEIR DIFFERENCE AND USES

The meaning of Freehand Drawing—How objects are made visible—
What a Perspective is—How a Perspective Drawing is obtained—
The use of a Perspective Drawing to the workman—An Orthographic
Projection, and how obtained—The meaning of Plan and Elevation   12—16

CHAPTER III

PRACTICAL GEOMETRY AND MECHANICAL DRAWING

The meaning of the term "Geometry"—The difference between Plane and
Solid Geometry—Definition of Geometrical terms used in the work—
Plane Geometrical Figures   ...   .   17—22

CHAPTER IV

PLANE GEOMETRY PROBLEMS

To divide a straight line into two equal parts—To erect a perpendicular
to a given straight line—To let fall a perpendicular to a straight
line—To bisect a given angle—To draw a line parallel to a given line
—To draw an angle equal to a given angle—To draw a line making
an angle with a given line   ...   .   23—27

viii   CONTENTS

PAGE

CHAPTER V
PLANE GEOMETRICAL FIGURES

To construct an equilateral triangle, an isosceles triangle, a scalene triangle
—To construct a square, a rectangle, a rhombus, a rhomboid, a trapezium, a regular pentagon, a hexagon, a regular octagon   ...   ...   28—34

CHAPTER VI
ORTHOGRAPHIC PROJECTION

The Planes of Projection—The difference between a vertical and a perpendicular plane—The relative position of the planes of projection—
The projections of a point and a straight line—The projections of a
line inclined to the planes of projection   ...   .   35—41

CHAPTER VII
PROJECTION OF PLANE FIGURES

The Projection of the Triangle—The square—The pentagon and the
hexagon, etc.   ...   .   42—46

CHAPTER VIII
THE PROJECTION OF SOLIDS

Definitions of the Plane Solids—The cube, the prism, the pyramid, etc.—
Models of the Solids necessary to the Student for a thorough knowledge of their projection—Elevations of objects given to find their
plans—Meaning of section, side elevation, sectional plans and eleva-
tions   ...   .   47—58

CHAPTER IX
PROJECTION IN THE UPPER PLANE

The front elevation given, to find the side elevation—The sectional elevations of the solid and hollow cube, prism and pyramid   ...   ...   59—70

CHAPTER X
PROJECTION FROM THE LOWER TO THE UPPER PLANE

The plans of objects given, to find their elevations and sectional elevations   71—76

CHAPTER XI
LINING-IN DRAWINGS IN INK

The kind of lines to be used—The direction in which the light is supposed
to fall on the object represented—To find the angle that the rays of
light make with the planes of projection—Why different qualities of
lines are used in Mechanical Drawing—The importance of correctness
in their application—How ink for lining-in a drawing should be
made—And how to fill the drawing-pen   ...   .   77—84

CONTENTS   ix

PAGE

CHAPTER XII
THE PROJECTION OF CURVED LINES

The definition of a curved line—The front elevation of a curved line being
given, to find its side elevation and plan—How to find the projections
of a line of double curvature—The projections of combined curved and
right lines—The plan of a circular plate being given, to find its elevation—To draw an ellipse ...   .   85—97

CHAPTER XIII
THE PROJECTION OF SOLIDS WITH CURVED SURFACES

The definitions of the cylinder, the cone, and the sphere—The plan of a
cylinder given, to find its elevation in various positions—The plan
of a cone given, to find its elevation in different positions ...   ...   98—102

CHAPTER XIV
THE PROJECTION OF THE SECTIONS OF A CYLINDER

The elevation of a cylinder given, to find its sectional elevation and
plan ...   .   103—105

CHAPTER XV
THE PROJECTION OF THE CONIC SECTIONS

The definitions of the sections of a cone—The plan and elevation of a cone
being given, to find its sectional projection—To find the true form of
any section of a cylinder or cone—The sections of a sphere and their
projections—Definitions of the subsidiary solids of revolution—The
lining-in in ink of solids with curved surfaces—How the light falls
upon them ...   .   106—119

CHAPTER XVI
THE PROJECTION OF OBJECTS INCLINED TO THE PLANES OF PROJECTION

To find the projection of a point, and line lying on an inclined plane—The
projection of plane figures when inclined to the planes of projection—
The projection of a solid when inclined to the planes of its projection
—The projections of the solid and hollow cube, the pyramid, and
cone, when inclined to the planes of their projection—The projections of a six-sided nut, when inclined to the planes of projection ...   120—141

CHAPTER XVII
THE PENETRATION AND INTERSECTION OF SOLIDS

The penetration of prisms by prisms at right angles to each other—The
penetration of prisms having their axes inclined to each other   ...   142—152

CONTENTS

CHAPTER XVIII

THE INTERSECTIONS OF PLANE SOLIDS (*continued*)

The penetration of a prism by a pyramid—The penetration of pyramids by pyramids ... ... 153—163

CHAPTER XIX

THE INTERSECTIONS OF SOLIDS HAVING CURVED SURFACES

The intersections of equal-sized cylinders at right angles to each other—The intersection of unequal-sized cylinders—The intersection of inclined cylinders—The intersection of the cylinder and cone—The intersection of cones by cylinders and cones—The intersection of the sphere by prisms and pyramids—The intersection of the sphere by the cone and cylinder ... ... 164—186

CHAPTER XX

THE DEVELOPMENT OF THE SURFACES OF SOLIDS

What a development means—What is a developable surface—The development of plane-surfaced solids—The development of the surface of a pyramidal-shaped solid, having a curved surface—The development of the oblique pyramid—The development of the surface of a right cylinder, and the frustum of a right cylinder—The development of the surface of an oblique cylinder, and of a right and oblique cone—The development of the surface of the sphere and hemisphere ... 187—211

INTRODUCTION

THE FIRST PRINCIPLES OF MECHANICAL AND ENGINEERING DRAWING

IT being incumbent on every one who aspires to become a really efficient Engineer, that he should possess a thorough practical knowledge of the Mechanical Draughtsman's art, we would in the outset of an attempt to explain the fundamental principles which govern its operations, observe, that the inducement to undertake such a task is the desire to place within the reach of every earnest engineering student and apprentice, a means of enabling him to *read* and to *make* such drawings as are placed before him in an engine factory to work from, and to prepare him for the subsequent study of engine and machine design.

It is assumed by the majority of engineering students and apprentices, that the drawing practised in the Drawing Office will be *taught* them upon their first admission to it, but an experience of many years in some of the principal offices in England, has made the writer alive to the fact, that so far as the "*principles*" which underlie the *practice* of the draughtsman's art are concerned, absolutely nothing is *taught* the student, and that if he ever acquires a knowledge of them, it will be by his own unaided study, independent of any drawing-office help. With a view, then, to the acquisition by the student of this all-important knowledge, in the best possible way, we have in the following pages formulated a method of imparting it, which from practical experience as a draughtsman, and teacher, we have found answers every requirement. Whether that method is an improvement on any now adopted, is left to those who earnestly follow its exposition to determine.

Before proceeding with that exposition, we would, however, put before the student, some facts bearing upon the study of drawing (and Mechanical Drawing more particularly), which may help him to appreciate the necessity that exists for his acquiring the ability to draw, if he desires to rise in his profession. Without wishing in the least to under-estimate the great worth of a really first-class skilled workman, who may have little or no knowledge of drawing, it is still

INTRODUCTION

a fact very generally admitted that just in proportion to the knowledge of drawing possessed by one workman over his fellow, so is he superior to him; and it follows that those ignorant of that art must hold a lower position as workmen, than those having a knowledge of it.

The utility of the power to draw may not present itself to the mind of the workman on its first suggestion to him, but a little thought about the matter will soon make it clear that it has a much closer connection with his daily work than he had any idea of. Neither spoken nor written language can at all times convey ideas that we wish to impart to another, and recourse must be had to some other means, more especially if those ideas relate to the *form* and *position* of material substances. To assist us in making our meaning clear, we must make use of what has been aptly called the "language of mechanics," or Drawing—a language which appeals at once to the eye for the truth of its assertions, and which enables us, without further assistance, to judge of the form, appearance, and dimensions of bodies.

To the intelligent mechanic, a real power of drawing is a priceless advantage, as it enables him to either reproduce a true representation of forms, that upon a casual inspection may have made an impression on his mind; or, on the other hand, to transfer to paper what he may have conceived, but which has not as yet had any existence. Many a valuable invention has been lost to posterity through the want of the power to draw, on the part of the would-be inventor.

Again, a knowledge of the graphic art is now demanded of all who are in any way connected with mechanical constructions of any kind, and no one can now hope to obtain any position of trust that an engineer fills, who has not acquired the power of correct drawing. It was long a fallacy with many, that draughtsmen were *born*, not made; that although a youth, or a man, may be taught to write—or copy letters—the law did not hold good as regarded drawing. This fallacy has happily gone the way of many others, and it is now held that those who will give to the study of Drawing the necessary concentration of thought, coupled with persistent effort, will undoubtedly attain its mastery and achieve success.

FIRST PRINCIPLES

OF

MECHANICAL AND ENGINEERING DRAWING

CHAPTER I

THE TOOLS AND MATERIALS REQUIRED BY THE STUDENT

1. Drawing-Board.—As all drawings of Mechanical and Engineering subjects are made on flat surfaces, and as the most suitable material on which such drawings may be made is paper, the first requisite of the student is a drawing-board, on which to lay or stretch the paper. The board should be made of well-seasoned pine, of a convenient size— say 23 in. by 17 in., which will take half-a-sheet of Imperial paper, leaving $\frac{1}{2}$ in. margin all round—$\frac{3}{4}$ in. in thickness, and fitted at the back, at right angles to its longest side, with a couple of hardwood battens, about 2 in. wide and $\frac{3}{4}$ in. thick; the use of these battens being to keep the board from casting or winding, and to allow of its expansion or contraction through changes of temperature. This latter purpose, however, is only effected by attaching the battens to the back of the board in the following manner :—At the middle of the length of each batten—which should be 1 in. less than the width of the board— a stout well-fitting wood screw is firmly inserted into it, and made to penetrate the board for about $\frac{1}{2}$ in., the head of the screw being made flush with the surface of the batten. On either side of this central screw two others, about $3\frac{1}{2}$ in. apart, are passed through oblong holes in the battens, and screwed into the body of the board until their heads are flush with the central one; fitted in this way the board itself can expand or contract lengthwise or crosswise, while its surface is prevented from warping or bending.

The *working* surface of the board—or its front side—should be perfectly smooth, but instead of being quite flat it should have a very slight camber, or rounding, breadthways, this latter feature in its construction being to prevent the possibility of a sheet of paper when stretched upon its surface having any vacuity beneath it. The four *edges* of the board need not form an *exact* rectangle, as much valuable time is often wasted in the attempt to produce such a board; but it

FIRST PRINCIPLES OF

will answer every purpose of the draughtsman so long as the *adjacent* edges at the lower left-hand corner of it are at right angles to each other, or square. To produce really good work in the shape of a mechanical drawing, *one* perfectly straight edge only is required on a drawing-board, and that the left one, which is always known as the *working* edge; but for the convenience of being able to draw a long line across the board at right angles to its lower edge, this edge is made truly square with that on the left side of the board.

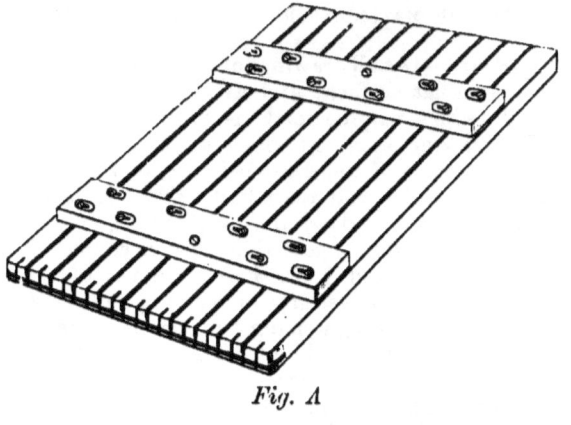

Fig. A

A further improvement in such a drawing-board as above described is made by cutting—lengthways—a series of narrow grooves in the back of it and inserting in its working edge a strip of ebony, to help in keeping it true, and to serve as a guide to the stock of the drawing square. Such an improved board is shown in Fig. A. There are other kinds of drawing-boards in use; but as the one described has stood the test of many years' service, and finds most favour in drawing offices, a detailed description of them is not necessary here. A reason for giving at such length a description of the kind of drawing-board so universally in use in modern engineering drawing offices is that it may be the means of inducing students and apprentices capable of handling joiners' or patternmakers' tools to make such a board for themselves, which, if made of good well-seasoned pine, free from knots and shakes, will retain its specially good features for years. Those, however, who may be unable to accomplish such a feat, may purchase such boards at a reasonable price from manufacturers of drawing materials, who make them a speciality.

2. Tee-Square.—The next most important adjunct to the drawing-board is the drawing- or tee-square. Some inexperienced youths, and even those of larger growth, have a notion that anything will do for a tee-square; but, if *correct* work is to be done, the tee-square is as important to the draughtsman as the drawing-board. It need not, however, be an expensive one, provided a knowledge of what constitutes a really serviceable and efficient tool is possessed by its intended user. As its

MECHANICAL AND ENGINEERING DRAWING

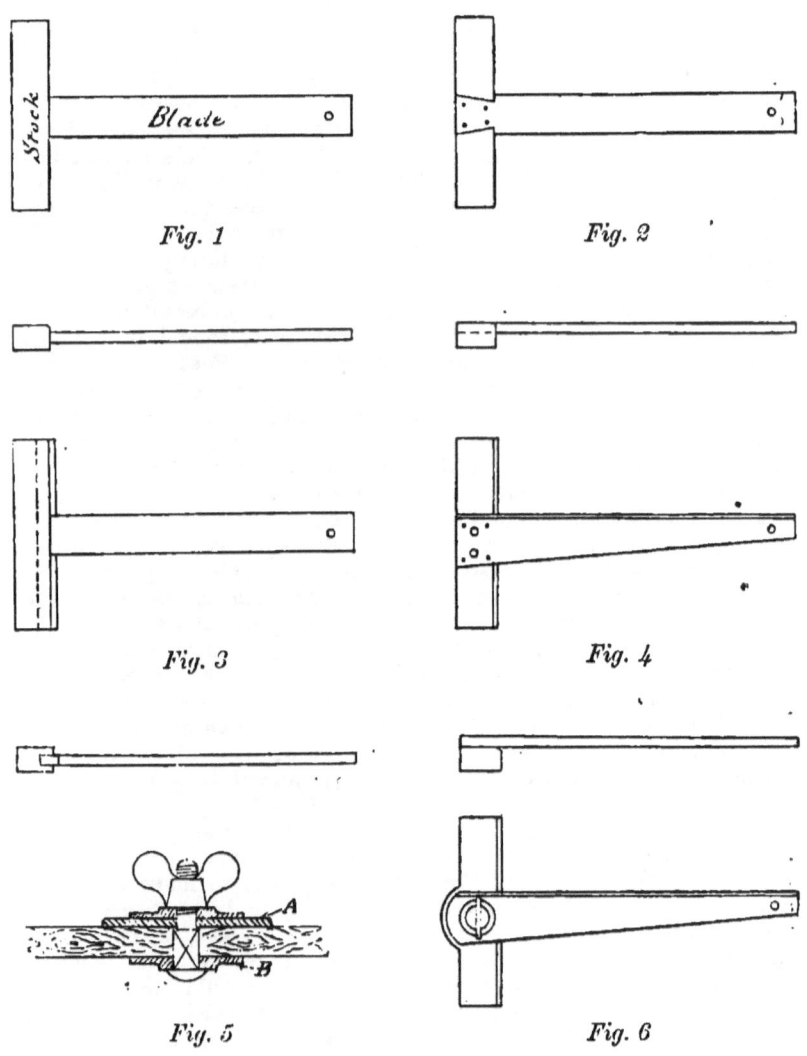

Fig. 1

Fig. 2

Fig. 3

Fig. 4

Fig. 5

Fig. 6

FIRST PRINCIPLES OF

name implies, it is an instrument in the form of the letter T, the two parts of it being known as stock and blade, the horizontal component of the letter being the *stock*, and the vertical one the *blade*. To form the square, the two parts are joined together in such a way as to make them exactly at right angles to each other; the stock, which is applied to the working edge of the drawing-board, being about one-third the length of the blade, and about three times its thickness.

The manner, however, in which the stock is united to the blade determines its adaptability or otherwise to the use made of it by the draughtsman. In some the stock is rectangular in section, and the blade morticed into it, as in Fig. 1. In others the blade is dovetailed and let into the stock for the whole of its thickness, as in Fig. 2; or morticed, as in Fig. 1, but fitted with a tongue-piece the length of the stock, as in Fig. 3. Neither of these plans is to be recommended; they involve unnecessary work and care in fitting during their manufacture, are more liable to get damaged in their usage, and are practically imperfect as a tee-square in some of its essential requirements. To be perfect in construction, a tee-square should be as light as is consistent with its necessary strength and stiffness of parts; it should be made of suitable material, easily manufactured, put together, and repaired, and withal as truly correct as it is possible to be made. Such a square is represented in Fig. 4; it has a taper blade, which is generally about double the width where secured to the stock as it is at the tip. Its tapering form serves two purposes, the primary one being that it adds strength and stiffness to the blade and prevents its buckling—a common fault with all parallel-bladed squares—and the other, its excess of width at the stock, prevents it from rocking, and gives ample room for securing it to the stock. The blade is also easily and correctly fitted to the stock, and has also one great advantage over all the others in that the set-squares used with it are far more easily manipulated than is possible with any of the three previously referred to.

3. In cases where many parallel lines have to be drawn, of lengths beyond the capabilities of ordinary set-squares, and in directions other than square with, or parallel to, the working edge of the drawing-board, it is convenient to have for use an *adjustable*-bladed tee-square, or one whose blade can be set at any desired angle. The blade of such a square should be tapered as in Fig. 4, but shaped at its wide end as in Fig. 6, and have a stock wide enough to allow for the surface required in the washers of the fittings necessary to make the blade adjustable. These fittings, though requiring to be well made and neatly finished, are not expensive or difficult to make, as they consist merely of two washers, a square-necked bolt, and a fly-nut, articles that any one capable of making a pair of calipers could supply himself with. Fig. 5 shows a section of these fittings, which are generally made in brass. The top and bottom washers A, B, are slightly dished on their faces to ensure contact with blade and stock, and the spread of the wings of the fly-nut is such as to give sufficient leverage for a good grip.

Reference is here made to an adjustable-bladed square, as one may possibly be required later on by the student; but there is no present necessity for the provision of such a tool, as all lines that may be required other than those drawn with the tee- and set-squares in conjunction, are easily put in by a proper manipulation of the set-squares, which will be explained in due course.

Set-Squares.—Of the *set-squares* used conjointly with the tee-square, those of 45° and 60° are all that are required by the student in the earlier stages of study. A 6-in. 45° and an 8-in. 60° set-squares are the most useful sizes. Framed ones, well made, of foreign manufacture, may now be obtained at a reasonable price, but the kind most generally in use are made of vulcanite. Those, however, of this material made with the middle part cut out to imitate framed wooden ones should be avoided, as they are very liable to fracture at the angles, and it is impossible to repair them.

The other requirements of the student of mechanical graphics, apart from what are known as *instruments*, are some pencils, drawing-pins, rubber, paper, and ink. A few words descriptive of the qualities that should obtain in each of these articles, that satisfactory work may be done, will be of advantage to him.

Pencils.—The present great demand for *pencils* has, notwithstanding the millions that are annually made and sold, added few to the number that are specially suited to the wants of the mechanical draughtsman. Many erroneously assume that any sort of pencil will suit a learner. No greater mistake can be made. If he is to acquire a draughtsman's habit of work, his first necessity will be a good, serviceable, reliable pencil—one that is neither too hard nor too soft, and that will retain a good point for a considerable time. The pencils now generally used in drawing offices are of Faber's make, which can be had of different degrees of hardness from H to six H's, the cedar covering of the lead being hexagonal in form, instead of round. But such pencils are too expensive for students' use. A good, serviceable pencil, made by Cohen, and known as the "Alexandra H pencil," has been in use by the writer for some years, and costs about half the price of Faber's. They are, however, of the ordinary round form, which is inimical to the draughtsman, it tending to cause them to be constantly rolling off his board and damaging their points. To obviate this, the writer's practice is to cut a flat side on the pencil throughout its whole length, taking care not to bend the pencil in doing this for fear of breaking the lead. If neatly done a perfectly flat side is produced, which serves as a guide to the way in which the pencil should be pointed and held, and will prevent any tendency to rolling, even if the drawing-board is much inclined. To do away, however, with the necessity for constantly sharpening the pencil, and thereby reducing its length at every such operation, pencils with movable leads have been in use in drawing offices for some years. They are far to be preferred, as the part of the pencil which is held by the fingers never alters in length, and the lead can be used to the last quarter of an inch. This kind of pencil is known as "Faber's artist's pencil," is hexagonal in outside form, and thus partly prevented from rolling. The acme of perfection in this class of pencil has, however, only lately been introduced, the part for holding the lead being triangular in

FIRST PRINCIPLES OF

section, which renders it easy to hold without turning in the fingers, and rolling off the drawing-board is impossible. It is made by Hardtmuth, of Vienna, but can be purchased of any photographic chemist or artists' colourman.

4. Drawing-Pins.—In the study of mechanical drawing in its

Fig. B

earlier stages, and even in the making of working drawings for shop use, it is not necessary or essential that the paper on which the drawing is made should be secured to the drawing-board in any other way than by pinning it. This is effected by the use of *drawing-pins*. There are, however, several kinds of drawing-pins to be had, and their variety is often the cause of difficulty in choosing, to the uninitiated user. A pin that would answer well the purpose of the free-hand draughtsman in putting a sheet of paper on his drawing-board, might be the very worst that a mechanical draughtsman could possibly use. The former, not needing a tee-square in the practice of his art, if he does not stretch his paper, pins it down to his board with any drawing-pins that are at hand. These may possibly be pins with heads a sixteenth of an inch thick, beautifully milled on their edges and perfectly flat on their under and upper sides. Such pins would be shunned by any mechanical draughtsman who wished to keep the edges of his tee- and set-squares intact and free from notches. Projecting the whole thickness of their heads from the surface of the paper, they would foul the edges of the tee- and set-squares and cause damage. The only kind of drawing-pin a mechanical draughtsman should use should have a head as thin as possible, without cutting at its edge, slightly concave on the under side or that next to the paper, and only so much convexity on its upper surface as will give it sufficient central thickness to enable the pin to be properly secured to it. There is neither sense nor reason in making the head of a drawing-pin half-an-inch in diameter if its circumferential edge does not bear on the paper when its pin is as far into the board as it will go. The purpose of the pin is to keep its head from rising from the surface of the paper, and it need only be long enough and strong enough to effect this. It is better practice to use four small, good-holding drawing-pins as shown in Fig. B, along the edge of a sheet of paper, than one large, clumsy, badly-made pin at each end of it. Suitable drawing-pins which answer every purpose required of them by the draughtsman are now to be obtained for half-a-crown per gross.

5. Paper.—As the student from the very commencement of learning to draw should study to acquire the good draughtsman's habits of work, and as one of these is the making of clear, sound lines in his drawing, whether in ink or pencil, it is advisable that he should accustom himself to draw on fairly-good paper. It is not meant by this that such paper as Whatman's is recommended for use in his preliminary work, but rather to guard him against purchasing soft, spongy paper, which will not stand the application of indiarubber for erasing, or of ink for lining in, without damaging its surface or causing the ink to run. Drawing-papers are of two kinds—viz., hand-made and machine-made. The former is the best, but is expensive; while the latter is made in great variety, and, as a consequence, of varying quality. Most students, in learning to draw, require a frequent use of the rubber; therefore a tolerably hard-faced paper is desirable. Since the advent of so much drawing as now obtains, a new special make of hard-faced, close-textured cartridge paper has been produced for students' use. It costs about 2d. per imperial sheet, and is very suitable for the purpose. For more advanced work there is, to the writer's knowledge, nothing that will compare with Whatman's smooth double-elephant paper, which takes the finest line either in ink or pencil.

Rubber.—For *cleaning* drawing-paper, a piece of soft, grey vulcanized rubber should be used, as it will not injure the surface of the paper if properly applied. Its only drawback is the appearance at times in it of small specks of some hard substance like coke-dust, which find their way into it during the process of manufacture; these, however, are easily removed when detected. For *erasing* any portion of a line in pencil, a piece of prepared white vulcanized rubber is the best—small rectangular pieces of this material are now to be had of any artists' colourman. What are called pencil-erasers, or rubber-sticks, are now in common use amongst draughtsmen for the same purpose. They are made in the form of a large square pencil, with rubber inserted in the body of it. To use it the wood is cut away, as is done in pointing an ordinary pencil, exposing the rubber, and it is then applied to the pencilled line with a to-and-fro motion of the hand, pressing lightly, until the line disappears.

Fig. C

Ink.—A further and all-important requisite to the student draughtsman is ink with which to line in his drawings after they have been carefully put in in pencil. We say this is an "important" requisite, because so much depends on its quality. It is generally known as India or China ink. The definition given of it in standard dictionaries— viz., "a substance made of lampblack and animal glue"—is no doubt answerable for the large amount of a material made and sold in Britain under its name. Pure India or China ink is only made in those countries, because the special wood from which it is produced is found only in those regions; therefore in purchasing ink for use on

8   FIRST PRINCIPLES OF

drawings, the only way to ensure its being the real article is to obtain it from a *bonâ fide* importer. The best mathematical instrument-makers are generally importers of it. It is sold in hexagonal sticks, as shown in Fig. C, and is expensive, but small oval and round sticks of it are to be had costing about a shilling each.

6. Before noticing the few "instruments" that are necessary when commencing the study of mechanical drawing, we think it advisable to show, in a combined sketch (Fig. D), the special tools—viz., drawing-board, tee- and set-squares—recommended, that the student may note the position they each should assume when in use. The tee-square should only be used in the two positions indicated by its outline in full and dotted lines. In the latter it will seldom be required. All lines at right angles to its edge, when in the first position, should be drawn with the 60° set-square applied, as shown. The 45° set-square is placed as it would be applied when a line is required at that angle near the left edge of the sheet of paper. Care should be taken, when drawing by lamp or gaslight, that the light is in such a position as to cause little or no shadow to be cast on the paper by the *edges* of the tee- or set-squares. This is important, as such shadows often cause errors in lining in, whether in ink or pencil. We may mention that the drawing-board and tee-square recommended for use are known as "Stanley's Improved," they having been introduced many years ago by Mr. W. F. Stanley, of London.

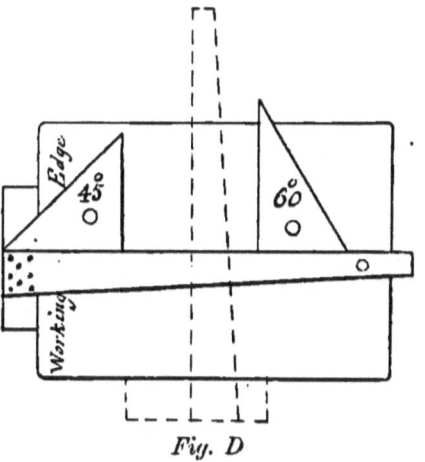

Fig. D

Instruments.—The *Drawing Instruments* required by the student draughtsman are few in number, and should be acquired as the necessity for their use arises. No greater mistake can be made than that of purchasing a "box of instruments," as it generally contains some articles that are never required, and is wanting in those that are necessary for the special kind of drawing practised. All that the student requires for use for some time is a pair of 6-in. compasses with a pen-and-pencil leg,

MECHANICAL AND ENGINEERING DRAWING 9

pen-and-pencil bows, a ruling- or drawing-pen, and a set of drawing-scales. For future service, everything depends on a proper choice in their purchase, more particularly if their use is to be continuous; and as we assume throughout this work that the student has little, if any, previous knowledge of the subject, it is especially necessary that he should know what constitutes a good serviceable instrument, as the possession of inferior ones will be a constant source of annoyance to him.

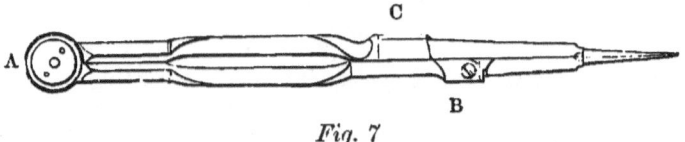

Fig. 7

7. In giving the characteristics of a *good* instrument, it is of the first importance to understand the use to which it is applied. With draughtsmen, a pair of *compasses* and a pair of *dividers* serve two very different purposes, and are therefore differently constructed, but their names and uses are often misunderstood. "Compasses" are never used for dividing, nor are "dividers" applicable to compass-work. Beginners should therefore note that the former are specially intended for putting in circular lines in pencil or ink, and that the proper and only use of the latter is the division or measuring-off of lines and spaces. These separate and distinct purposes give at once a clue to their proper form and construction. They are both instruments with two movable legs, joined together by a forked end, and secured by a pin and washer, as shown in Figs. 7 and 8 at A, A. The compasses, however, being used to draw circular lines, or lines described about a point everywhere equi-distant from it, should have jointed legs, one with a knee-joint at B, and the other with a socket, as at C, to enable it to be easily removed and replaced by the ink- or pencil-points D, E, Fig. 10, when required. The purpose of the knee-joints shown at B in the compasses, and *b b* in the pen and pencil points, is to enable the lower parts attached to them to be adjusted perpendicular to the surface of the paper, in order to obtain a truly circular line, and to allow both nibs of the inking-point to bear fairly upon it.

Dividers, which are not necessary to the student for some time forward in his study, should have legs of equal length, but without joints, as in Fig. 8, their lower parts being made of steel of triangular section to within $\frac{3}{4}$ in. of the ends, which should be gradually worked off into nicely-rounded points, as shown. This latter feature is one that

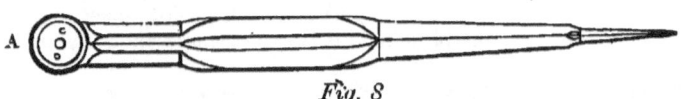

Fig. 8

should obtain in the points of compasses, bows, etc. *Triangular*-pointed instruments should never be used, as their points act the part of a

10   FIRST PRINCIPLES OF

rimer, cutting their way through the paper into the drawing-board, making unsightly holes, and causing them to describe anything but true circles.

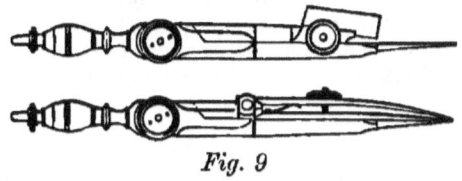

Fig. 9

Pen-and-pencil Bows are compasses intended for putting in smaller circles and circular arcs. Single-jointed ones, such as are shown in Fig. 9, will serve all the present wants of the student, if well made. The socket in the pencil-bow should be tubular, and of a size to take leads, and not lead-pencils. As these two instruments will be much oftener used than any other, it is advisable that the student should supply himself with the best to be afforded, as they will amply repay any present outlay.

What are known as "half sets," shown in Fig. 10, are now specially made by drawing-instrument makers, for the use of students. They comprise compasses, lengthening bar, pen and pencil point, and knife key, and are a very serviceable outfit if well made.

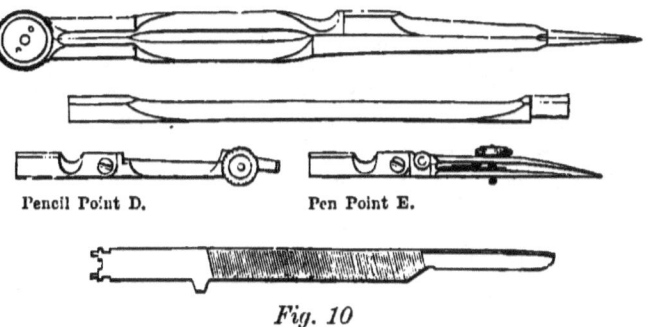

Pencil Point D.   Pen Point E.

Fig. 10

In selecting the foregoing instruments, care should be taken that they are all sector-jointed with double-leaves and well made; there should be no shake or slackness in any of them, and they should be equally stiff in the joints at any point from being full open to closing. The test for a pair of compasses is to open out their legs well apart and then to fold each lower half-leg together—if the points meet each other truly, they are correct in the joints; if they cross one another, the joints are not properly made.

Drawing or Ruling-pens are of two kinds—viz., those made with a jointed nib, as in Fig. 11, and those without a joint, as in Fig. 11A. The former, though more expensive, is to be preferred, on account of the facility in cleaning and sharpening; but the latter is a very serviceable

MECHANICAL AND ENGINEERING DRAWING 11

pen, if well made and finished. It will be observed in the sketch of the first, that the under or fixed nib is much straighter and thicker than the hinged one; this is so made to resist the pressure of the hand upon it when drawn along the edge of the tee- or set-squares. In all ordinary pens the nibs are of equal thickness, and the hand-pressure tends to close them and prevent the flow of the ink; but by providing a stout springless inner nib this tendency is overcome. The stem or handle of this pen, it will be noticed, is squared, to indicate how it should be held by the fingers when in use.

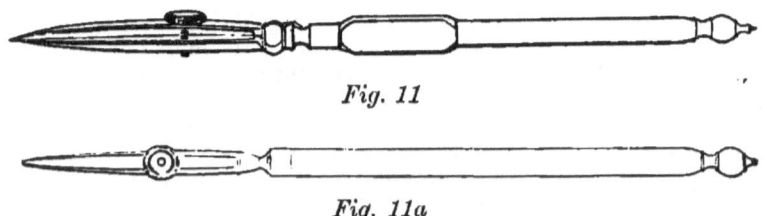

Fig. 11

Fig. 11a

The Drawing-scales recommended for present use by the student are a set of three lately introduced by Messrs. Jackson Bros., of Leeds, made of pliable varnished beech, and giving twelve scales of the standard units of measurement generally used in engineering drawing. They are decidedly to be preferred to any cardboard-scale, as the dividing is well done and there is no tendency to double up or get dirty by use. When the student acquires a more perfect knowledge of the use of instruments and scales, he can add to his stock already in possession whatever is necessary, always bearing in mind that a good tool in the hands of one who knows how to use it will invariably do better work, and is to be preferred to one of inferior quality; in the meantime, those herein recommended are all-sufficient for present requirements.

CHAPTER II

MECHANICAL AND FREEHAND DRAWING : THEIR DIFFERENCE AND USES

8. BEFORE proceeding with an exposition of the principles on which the practice of mechanical drawing is based, it is necessary that the student—who is assumed to have no previous knowledge of the subject—should thoroughly understand the radical difference, in character and application, which exists between it and that kind of drawing known as " freehand."

The generic term " drawing," strictly speaking, is the art of representing objects on a surface—generally flat—by means of *lines* showing their forms and general contour, independent of colour or shading ; for the latter, without form, would be meaningless and incapable of expressing anything. Freehand drawing is the practice of the art of drawing by means of the hand, the eye alone controlling and guiding the tool or instrument used for delineation. The hand guided by the eye can, however, only picture or draw what is seen from *one position* at a time ; for were it otherwise, a distorted view of the object would be the result, as its *appearance* to the eye from one point of view would be different to that from any other.

All objects are made visible to the sense of seeing by the agency of *light*, whether natural or artificial, for without light it would be impossible to distinguish one object from another. To the artist or draughtsman, light is a stream of matter given off by a luminous body, travelling from its source in thin straight lines—or *rays*—to the object illumined, from which it is reflected or transmitted in the same way to his eye. What is seen, or is apparent to his sense of sight, he depicts or draws on his paper. If he changes his position with respect to the illumined object, he sees it differently, and obtains a different view of it ; each such view, if correctly drawn, is known as a " perspective," and would agree with that obtained in the following manner.

In the diagram (Fig. 12) let HP represent a flat surface, such as a piece of ground or a floor, exposed to sunlight, and VP a sheet of glass set up on HP, in a vertical position. At any distance to the left of VP, and parallel to it, is erected a piece of fencing OO, having its top and bottom edges parallel to HP, and its side edges

12

MECHANICAL AND ENGINEERING DRAWING

13

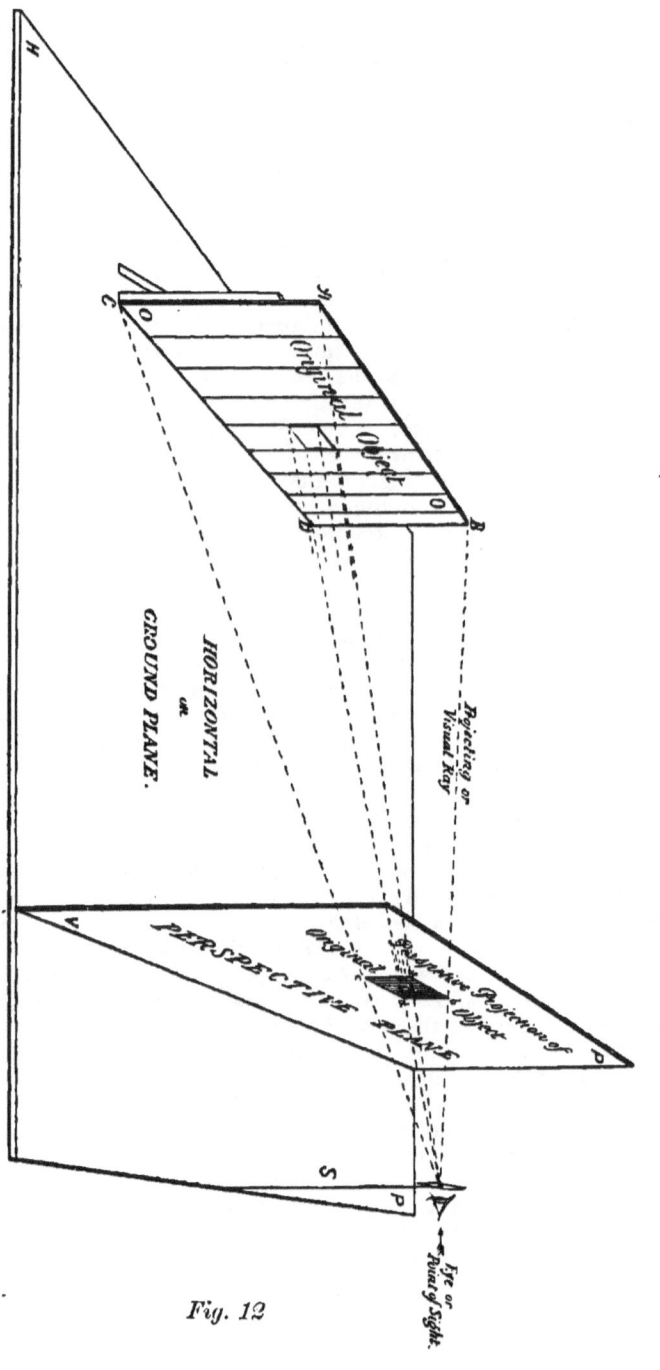

Fig. 12

14   FIRST PRINCIPLES OF

perpendicular to it. At a given distance to the right of VP, and perpendicular to HP, a staff S, surmounted by a small rectangular plate of any opaque material, and pierced with a sight-hole is fixed; the height of this sight-hole from HP being supposed to equal that of an observer's eye from the ground. The sheet of glass VP being transparent, it is evident that the spectator, on looking through the sight-hole, will see the whole of the piece of fencing, and can judge of its *appearance* from the *position* occupied by his eye. If he wish for a record of this appearance he can obtain it by drawing on the glass what he sees through the sight-hole. The view he would get would be a perspective of the original object OO, or the fence. But its contour or outline on the glass, although similar, would be much smaller than its original. How much smaller, would entirely depend upon the distance between the eye at sight-hole, the sheet of glass VP, and the fencing OO. It is evident that the nearer VP is to OO, the eye remaining in the same position, the larger would be its image or picture upon VP, and the converse of this would obtain were the conditions reversed.

It will be seen from the diagram that the perspective view of the original object is obtained by finding where the luminous or visual rays —represented by broken lines—proceeding from its principal points, are intercepted on VP in their passage to the eye, and then joining such points by right lines as in the original. Now, as these visual rays, or "projectors," are the means by which the view of the object is *projected* or thrown on VP, such a view is called a "projection," and in the special case we are considering a "perspective projection." In such a delineation it is apparent that all rays proceeding from the visual points in the object form a pyramid, the vertex of which is the point where they meet in the eye; and from this fact it will at once be seen that a perspective drawing of an object can serve no other practical purpose than that of showing its appearance when viewed from a certain fixed position, for its boundary lines altering with the altered position of the spectator, it is difficult to determine their *actual lengths*, as they only bear a relative proportion to their originals. As they cannot be measured with an ordinary rule or scale, it would be impossible to construct a machine or erect a building from such drawings. In perspective drawing, HP in the diagram is known as the horizontal or *ground plane*, and VP the perspective or *picture plane*, which latter is always supposed to be transparent, although actually represented by the artist's sheet of drawing-paper or canvas.

9. As, then, a perspective, or freehand drawing, does not fulfil the requirements of the workman, in that he cannot determine at sight the actual form, dimensions, or arrangement of the piece of work he is called upon to execute, some other method of delineation becomes a necessity. This want is supplied in what is generally known as a "mechanical drawing," or a drawing obtained by the correct application of the principles of a kind of projection called "orthographic," which gives results differing widely from that already explained, in that it affords a means of at once determining the actual form, size, and disposition of every part of the object represented, and gives an adept in the application of its principles the power to commit to paper the entire design of a machine or engine, that will enable the engineer or machinist to determine at sight whether the stationary and working parts of one or the other are disposed in such a way as to meet the requirements for which they were designed. In fact, a mechanical drawing is the only efficient way of describing by means of *lines*, properly disposed according to fixed rules, the actual construction and arrangement of a piece of mechanism.

As "orthographic" projection is the basis of mechanical and engineering drawing, its difference as compared with perspective projection must be understood before the study and application of its principles are entered npon. An important consideration in connection with either kind of projection is, that the bodies, or objects, whose forms it is wished to depict on paper, are in all cases assumed to be illumined by solar light, and have the power of reflecting or throwing off the light that is cast upon them. As the source of light—or the sun—is at a comparatively *infinite* distance from all objects illumined by it, its rays will not sensibly diverge, or approach each other, but may be regarded as exactly parallel among themselves. Then if, instead of the rays from an illumined object being reflected so as to converge in the eye—as in perspective projection—they be conceived as travelling from the object in *parallel* lines, till intercepted on a plane surface at right angles to themselves, and the points of interception be joined by straight or curved lines, the representation thus formed on that surface will be an "orthographic" projection of the original object. In this case the visual or projecting rays, being always parallel to each other and perpendicular to the surface on which they are projected, form a prism; and it follows, that, however far that surface is from the object, its representation remains the same, and the projected length of all its lines parallel to that surface will be of the same length as in the orginal, and therefore their exact dimensions can be at once ascertained.

It will be understood from this explanation that instead of the eye being stationary and viewing the object from one point alone, as in perspective, it is in orthographic projection supposed to move in such a way as to be directly opposite to each of the principal points of the object, the projecting rays from it being always perpendicular to the plane on which its image is projected. It is manifest, however, that in this way only one projection of an object is obtained; but as any solid body has more than one dimension, it becomes evident that more than one view of it must be given before its other dimensions can be ascertained. To this end it is usual to determine its projections on two planes, which are always at right angles to each other, and from these correct and definite ideas as to its shape and dimensions may at once be obtained.

10. To illustrate the foregoing diagrammatically, let HP (Fig. 13) be a horizontal plane, and VP another plane at right angles or perpendicular to HP. At any distance from VP, and in front of it, a rod R is set up perpendicular to HP, supporting on its upper end a bar F of rectangular section and a given length. Visual rays or

16 MECHANICAL AND ENGINEERING DRAWING

projectors parallel among themselves and perpendicular to VP are shown proceeding from the corners A, B, C, D, of the bar penetrating VP in *a, b, c, d.* As the edges of the original object, or the rectangular bar F, are all straight, it follows that if *a, b, c, d* on VP be joined by straight lines, an orthographic projection of the face A, B, C, D, of the bar will have been obtained, which will, on measurement, be found to be an exact counterpart of it. But this projection only gives the *length* and *depth* of the bar; and as it is necessary to know its other dimension, or *width*, a view showing that dimension must be obtained. Now it is evident that a view of the bar, looking at it from above and in the direction of the arrow, will

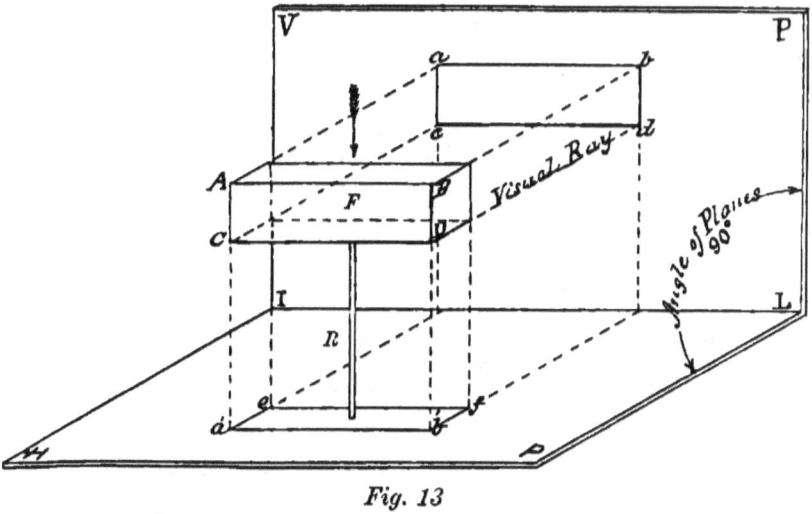

Fig. 13

supply the information required. If, then, visual rays, or projectors, proceed as before from the four corners of the face of the bar seen from above, to the plane HP below, they will penetrate that plane at the points *a', b', e, f,* and these points being joined as before—as the same conditions obtain—there is produced on HP an orthographic projection of the top face of the bar which determines its width. With these two projections, or views of the original, it will be seen that a workman could produce any number of such bars without the assistance of a model or other guide. To distinguish the two projections of the same object, the one obtained on VP is known as an "elevation" or *vertical* projection, and that obtained on HP is called a "plan" or horizontal projection.

CHAPTER III

PRACTICAL GEOMETRY AND MECHANICAL DRAWING

As it has been necessary, in explaining the difference between a mechanical and a freehand or perspective drawing, to make use of terms which pre-suppose a knowledge of geometry by the student which he may not possess, and as it is advisable to take nothing for granted in the exposition of our subject, it will be necessary at this stage to define the meaning of the geometrical terms that will be made use of as we proceed, and to show how, by a special combination of lines, those geometrical figures are constructed which form the surfaces of objects whose delineations are subsequently to be obtained by orthographic projection.

The term "geometry," in its generally-accepted sense, means the science or knowledge of *magnitude* reduced to system, and has to do with the measurement of lines, surfaces, and solids. It has, like other sciences, two sides or branches, one "theoretical," which demonstrates or proves its principles, and the other "practical," or that which applies those principles to construction. Theoretical geometry, or Euclid, will seldom be referred to in the course of this work, as most of the demonstrations used are self-evident; but practical geometry—a sub-divison of which is the basis of our subject—must be understood by the student to such an extent as will enable him to work out the problems that will arise in the exposition of it.

The two parts into which practical geometry is divided are : Plane geometry, which has reference only to the solution of questions relating to points, lines, and figures, situated in *one plane ;* and solid geometry, which shows by special representations on *two or more* defined planes, the relations of the points, lines, and surfaces of bodies having length, breadth, and thickness.

We would, in passing, guard the student, on his entering on the study of geometrical drawing, against wasting valuable time in working out the problems—many of which will be of no use to him—given in most text-books on the two subjects of plane and solid geometry, as all that is absolutely necessary for him to know in connection with either will be explained to him as occasion arises.

As we cannot form a conception of the *magnitude* of any material

FIRST PRINCIPLES OF

object without reference to one or more of its dimensions, and as each of these involves the idea of *extension* in some direction, the word *length*, or its representative, "a line," would appear to be the first *term* used in geometry requiring definition, but as a *line* can have no existence till it is generated or drawn, our first term must be that of the *generator*, or "point." We therefore define—

A *point*, as having no magnitude, that it is used to denote "position" only, and is represented geometrically by a dot or mark made by any pointed instrument, such as a pen, pencil, etc.

A *line*, as the path made by a point moving over a surface. It may be straight, crooked, or curved, according to the direction in which the point travels or moves.

A *straight line*, as the shortest path that can be made by a point moving from one position to another, or the nearest distance between two points, as the line A between points 1 and 2.

A *crooked line*, as the path of a point that has changed its direction after moving in a straight line for a given distance—1 to 2,—as the line B from 1 to 3.

A *curved line*, as the path of a point that continually changes its direction, as the line C from 4 to 5.

If the path of a moving point changes in such a way as to enclose a certain amount of surface, then the enclosed surface is called a "figure," and the path its *boundary line*, as in Fig. 14, where the "point path" from *a*, through *b*, *c*, *d*, defines the form of the figure.

If a point move continuously in such a way as to be always at a given distance from some *fixed* point, then the surface enclosed by the "point path" becomes the figure called a "circle," as the continuous line ABC (Fig. 15), any point in which is equi-distant from D, which is called its *centre*.

It is evident from the foregoing that two *straight* lines cannot *enclose* a surface, or *form* a figure, but that one such line in combination with a curved or a crooked line will effect this, as shown in Figs. 16 and 17, where we have in one case a straight line and a crooked one, and in the other a straight and a curved line, combined to form figures.

A *surface* is a magnitude that has extension in two directions only —viz., lengthwise and crosswise. Its dimensions—with one or two exceptions—are given as *length* and *breadth*.

A *plane surface* is one that a perfectly straight edge will touch or coincide with if applied to it in any direction. A mathematical or perfectly true plane does not exist—it can only be imagined.

Parallel straight lines are the point paths of two lines on a *plane* surface that are everywhere equi-distant from one another, as the lines A and B.

Converging straight lines are the point paths of two lines on a plane surface, which, if continued, meet and cross each other as the lines C, D. When the paths or lines *increase* their distance from each other as they leave the meeting point, they are said to *diverge*.

An *angle* is formed when two straight lines meet each other in a point, as D meets F in *d* (Fig. 18). If the inclination of one line to the other be such that the angles are equal on both sides of the meeting

MECHANICAL AND ENGINEERING DRAWING   19

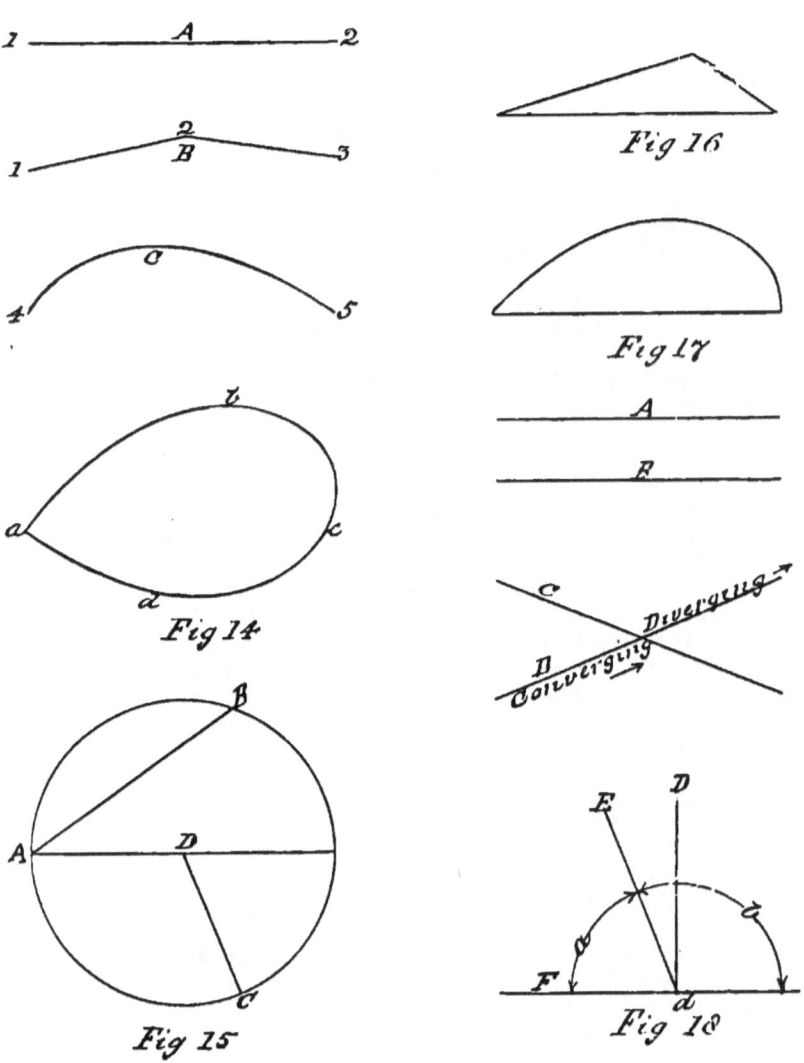

Fig 16

Fig 17

Fig 14

Fig 15

Fig 18

point, then the angle formed by the lines D and F is a *right angle*. If they are not equal, as in the meeting of E and F in *d*, then the *smaller* of the two, or angle *a*, will be an *acute* angle, and the *larger*, or angle *b*, will be an *obtuse* angle. And as the line D makes equal angles on both sides of it with the line F, the two lines D and F are *perpendicular* to each other.

11. As angles cannot be *measured* without a knowledge of the parts and divisions of the circle (Fig. 15), we must, before giving further definitions of plane figures, explain these. The boundary line ABC of this figure is called its *circumference*. Any straight line drawn through D, its *centre*, and touching the circumference on both sides, is a *diameter*. Half of such a line is called a *radius*. Any portion of the circumference, such as from A to B, would be an *arc*, and a straight line joining A and B the *chord* of that arc; the space enclosed by the arc AB and its chord is called a *segment*, and that by the arc AC and the two radii AD, CD, a *sector*. One quarter of the whole figure or circle is a *quadrant*, and one half of it a *semi-circle*. For the measurement of angles, arcs, chords, etc., the circumference of every circle is supposed to be divided into 360 equal parts called *degrees*, which are indicated when speaking of them by attaching a small circle to the right of the number stated—as 30°, 60°, etc. (or 30 degrees, 60 degrees, etc.). A quadrant, therefore, contains 90 degrees, and a semi-circle 180 degrees. The size of any angle is determined by the number of degrees contained in the arc subtending the angle, described about the angular point *d*, as a centre (Fig. 18). The *complement* of the angle *a* is the number of degrees it is wanting to make it a right angle, and its *supplement* is the number of degrees contained in the angle *b*.

Circles are *concentric* when they have the same centre as in Fig. 19, and *eccentric* when their centres are different, as in Fig. 20.

A *tangent* is a straight line which touches a circle or a curve in one point, and when produced does not cut it, as in Fig. 21.

Tangent circles and *tangent curves* are those which touch each other in one point, but do not cut as in Fig. 22.

The *point of contact* is that point where a tangent touches a circle or a curve, or where two tangent curves touch, as *a* in Fig. 22.

11a. In continuing our definitions of plane figures, we take first those constructed with the least number of *straight* lines that will enclose a space, which is *three*. Such figures are called triangles or three-angled, and are named according to the disposition of their sides and quality of their angles.

An *equilateral* triangle has equal sides and equal angles, which are all acute, as in Fig. 23.

An *isosceles* triangle has two sides equal and two of its angles always acute, the third angle being acute or obtuse, dependent on the length of its third side, as Figs. 24 and 25, the latter having one obtuse angle at *a*.

A *right-angled* triangle has one of its angles a right angle, the other two being acute, as Fig. 26.

A *scalene* triangle has three unequal sides, as Fig. 27; an *obtuse-angled* triangle has one obtuse angle, as Fig. 28; and an *acute-angled* triangle has all its angles acute.

MECHANICAL AND ENGINEERING DRAWING

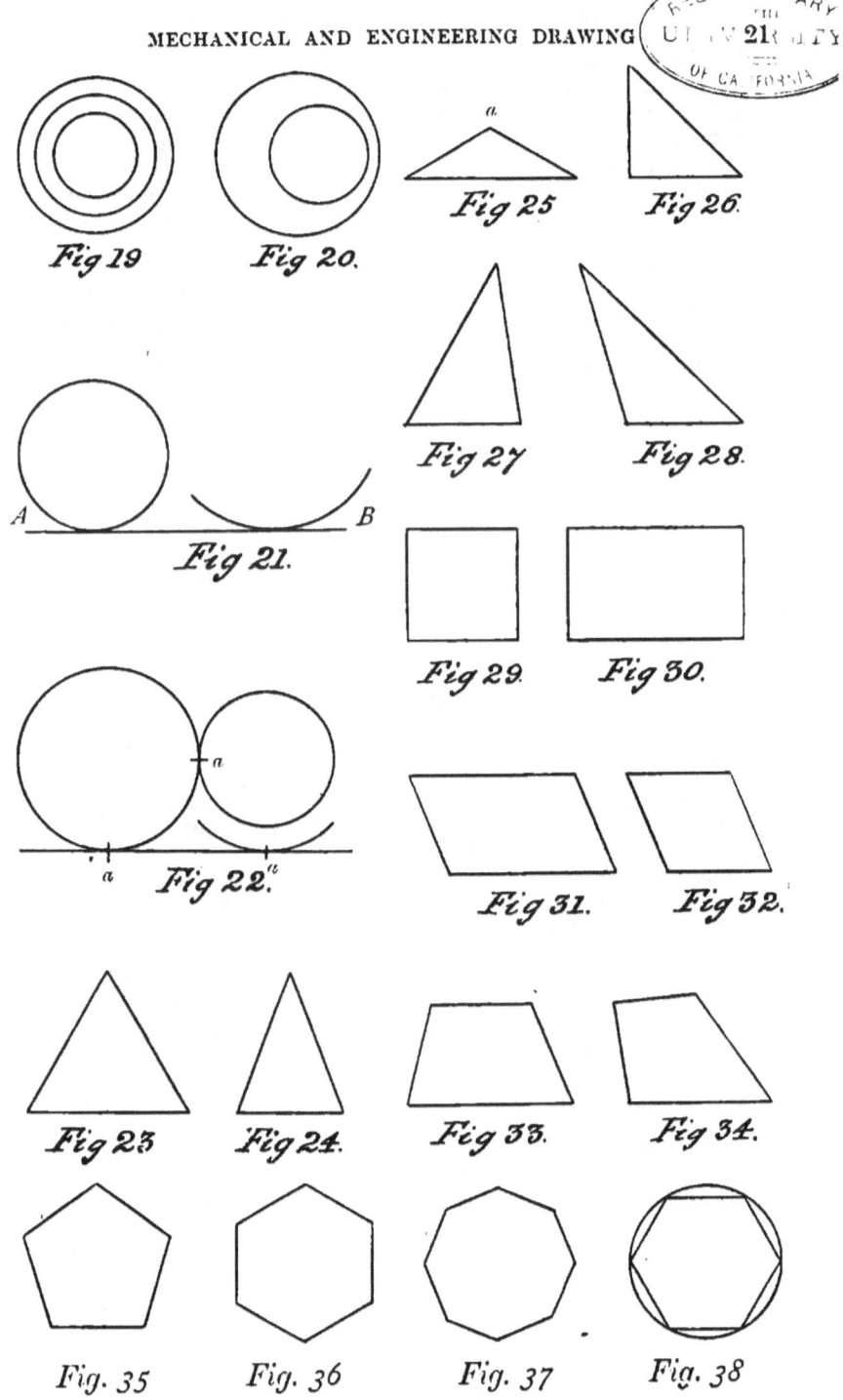

Fig 19

Fig 20.

Fig 25

Fig 26.

Fig 21.

Fig 27

Fig 28.

Fig 22.

Fig 29.

Fig 30.

Fig 31.

Fig 32.

Fig 23

Fig 24.

Fig 33.

Fig 34.

Fig. 35

Fig. 36

Fig. 37

Fig. 38

22   MECHANICAL AND ENGINEERING DRAWING

Of four-sided figures bounded by straight lines—

A *square* has all its sides equal, and its angles right angles, as in Fig. 29. A *rectangle* has opposite sides pairs, and parallel, and its angles right angles, as in Fig. 30. A *parallelogram*, or *rhomboid*, has two pairs of parallel sides, as Fig. 31. A *rhombus* has all its sides equal, two of its angles being acute, as Fig. 32. A *trapezoid* has only two sides parallel, as Fig. 33. A *trapezium* is an irregular figure of four sides, none of which are parallel, as Fig. 34.

A *regular polygon* is a figure having all its sides and angles equal. One of five sides is a *pentagon*, as Fig. 35. One of six, a *hexagon*, as Fig. 36. One of eight sides, an *octagon*, as Fig. 37. An *irregular polygon* is a figure whose sides and angles are unequal.

The *centre* of a polygon is a point that is equi-distant from its sides and its angular points. A polygon is *circumscribed* when all its angular points touch a circle described about it, as in Fig. 38.

A circle is *inscribed* in a polygon when its circumference touches all the sides of the polygon, as in Fig. 39.

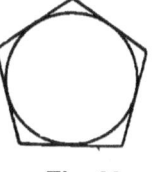   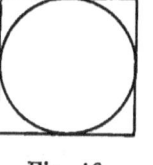   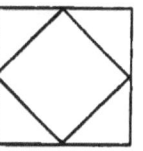

Fig. 39   *Fig. 40*   *Fig. 41*

A right or straight-lined figure is *described about* a circle when all the sides of the figure touch the circumference of the circle, as in Fig. 40; and a straight-lined figure is *inscribed* in another such figure when the *angular* points of the inscribed figure are upon the sides of the figure in which it is inscribed, as in Fig. 41.

A straight line joining the *opposite* angular points of any four-sided figure is a *diagonal*. A square or a rectangle has its diagonals of equal length. In a rhombus, rhomboid, trapezium, and trapezoid the diagonals are unequal.

As an apprenticeship to any mechanical trade cannot be served in a factory or workshop without the apprentice, as he advances in knowledge and skill in it, being often called upon to *line out* his own work, it is necessary that he should be able to draw any of the above-described plane geometrical figures *on the material* he works in, with a straight-edge, a scriber, and a pair of shop compasses, instead of the tee- and set-squares, etc., of the draughtsman. With this fact in view, we shall give in the solution of each of the following problems the simplest possible method of construction, it being undesirable to burden the memory of the student with the many ways of solving them to be found in text-books.

CHAPTER IV

PLANE GEOMETRY PROBLEMS

12. As the lines forming the *boundaries* and determining the *forms* of the plane geometrical figures previously described, have a certain relative position, it is necessary, before attempting to construct the figures themselves, that we know how to draw geometrically, lines having any defined relation to each other. As this knowledge is generally imparted in the form of *problems*, with their solutions, we shall adopt the same plan; but in explaining the constructions do not confine ourselves to any orthodox method where a simpler one may be used. The student will remember that in solving the subsequent problems, only the tools mentioned in the last paragraph of the previous chapter are to be used, as the assistance of either a drawing-board or squares is inadmissible. As it is not always possible to apply a rule or a scale to a line, when we wish to sub-divide it into parts, our first problem is—

Problem 1 (Fig. 42).—*To divide a given straight line into two equal parts.*

Now, if the given line is *near the edge* of the material on which it is drawn, a different method of construction must be used to that which would be possible if the line were *some distance from* that edge. In the former case proceed as follows :—With a distance greater than half the length of the given line AB, as a radius, and with A and B as centres, describe arcs cutting each other in C, and with a still larger radius than before, and from the *same* centres describe arcs cutting each other in D ; then the point E, where a straight-edge laid exactly on C and D crosses the line AB, is the middle of the given line, and divides it into two equal parts. In the latter case, with the distance greater than half AB (which may be gauged by eye) as radius, and from A and B as centres, describe arcs on *both* sides of the line AB, cutting each other in C and F. Then the straight-edge applied to C and F will give E in AB as its point of section, dividing it into two parts of equal length.

23

Problem 2 (Fig. 43).—*At a given point C, in a straight line, to erect a perpendicular.*

Here the point may be near the *middle* of the line or near the *end* of it. If the former, as at C, in the line AB, and if AB be near the edge of the material, proceed as follows :—Set off from C, on either side of it, equal distances, as CD, CE, and from D and E as centres, with a radius greater than half the distance between D and E, draw arcs cutting each other in F, then a line drawn through F and C will be perpendicular to AB. If the given point is near the *end* of the line and the edge of the material, as A in BD (Fig. 44), then from any point *a*, above BD, and with a radius equal to *a* A, describe an arc CAT, passing through A, and cutting BD in T. Draw a line from T through *a*, and produce it till it cuts the arc in C. A line from C through A will be perpendicular to BD at A.

Problem 3 (Fig. 45).—*From a given point A, above a straight line BC, to let fall a perpendicular to that line.*

Here the point may be nearly over the middle, or over the end of the given line. If in the first position, with any radius greater than the distance from the point A to the line BC, describe an arc cutting BC in D and E, and from points D and E as centres, with a radius greater than half the distance between D and E, draw arcs cutting each other in *a* and *b ;* then a line drawn through the given point A and the intersections of the arcs in *a* and *b* will be the required perpendicular. If the point is nearly *over the end* of the given line, as *b* in Fig. 46 is over AB, from *b*, draw a line intersecting AB in C, and bisect it in S; with SC as radius and S as centre, describe an arc cutting AB in D, join *b* and D, and the line will be perpendicular to AB. The student will notice that the construction in the second cases of Problems 2 and 3 is similar. This arises from the fact that the line drawn to the given point has in each case to be at right angles to the given line, and as the angle in a semi-circle is always a right angle, the problem is to draw a semi-circle that shall contain the three angular points of a right-angled triangle, one of which is the given point in the problem.

Problem 4 (Fig. 47).—*To bisect (or divide into two equal parts) a given angle.*

When speaking of an angle, it is usual to name it by affixing either a single letter at the angular point, or a letter to each of its lines and the angular point, the one denoting the latter being always the second. In the problem, let BAC be the given angle. With any convenient radius set off from A equal distances on BA and CA in the points D and E, and from these points, with a radius greater than half the distance across from D to E, draw arcs intersecting in F ; a line through F and A will bisect the angle BAC. This construction, it

MECHANICAL AND ENGINEERING DRAWING   25

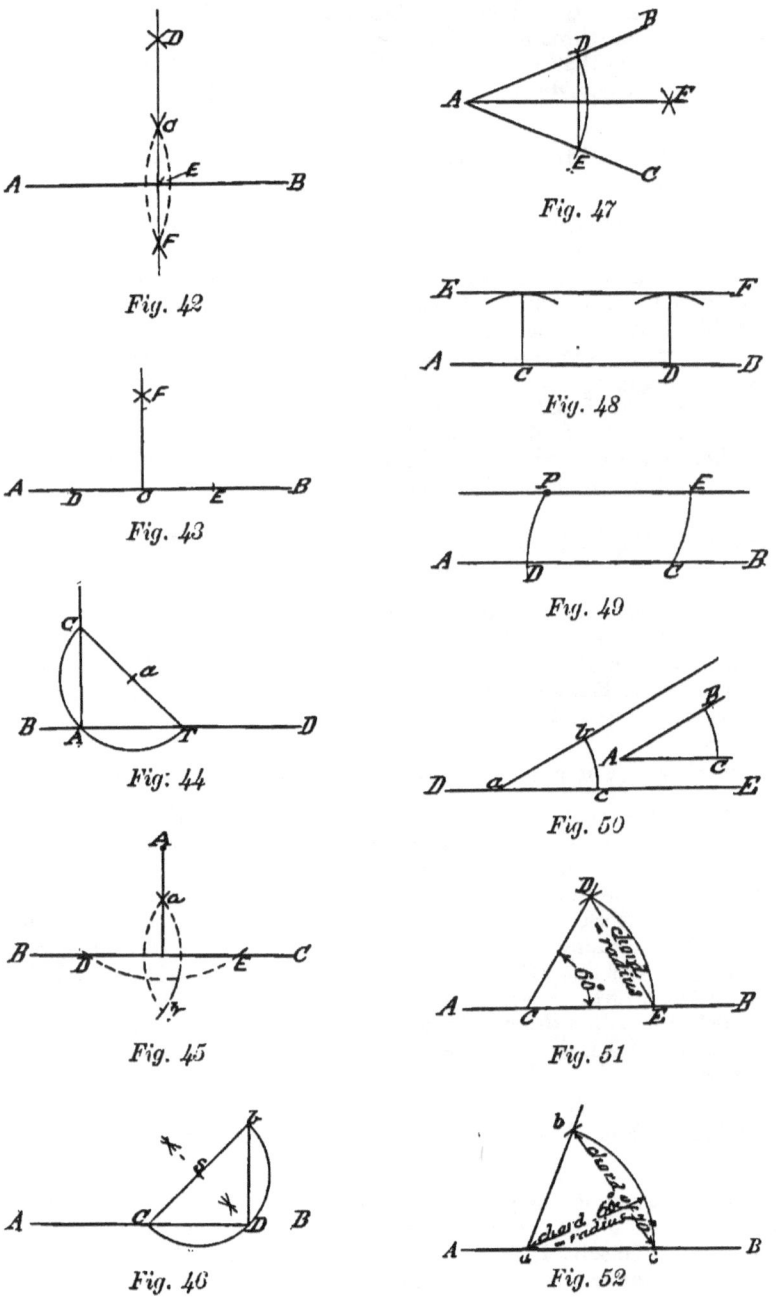

Fig. 42

Fig. 43

Fig. 44

Fig. 45

Fig. 46

Fig. 47

Fig. 48

Fig. 49

Fig. 50

Fig. 51

Fig. 52

will be seen, is tantamount to bisecting a line from D to E, and drawing a line through its bisection and point A, the only requisite condition being that the two points D and E in the lines forming the angle must be equi-distant from the angular point A.

Problem 5 (Fig. 48).—*To draw a line parallel to a given line at a given distance from it.*

Here it is evident that if from any two points C and D in the given line AB, arcs be drawn, of a radius equal to the given distance the two lines are to be apart, and a line EF be drawn tangent to those arcs, then the line EF will be parallel to the given line AB. This is the simplest possible solution of the problem, involving the least work, but requires care in drawing the parallel line exactly tangent to the arcs. Another solution, requiring much more work in the construction, is the following : —At the points C and D, in line AB (Fig. 48), erect two perpendiculars to AB, and set off on each of them from C and D the distance the parallel lines are to be apart. Through the two points obtained draw a line, and it will be parallel to the given line AB.

Problem 6 (Fig. 49).—*Through a point P, to draw a line parallel to a given line AB.*

With P as a centre and any convenient radius, describe an arc EC, cutting the given line AB in C, and from C as a centre, with the same radius, draw an arc through P, cutting AB in D. Set off the distance PD on the arc EC, and through P and E draw a line ; it will be parallel to the given line A B.

Problem 7 (Fig. 50).—*To draw an angle equal to a given angle A.*

This means that two lines are to be drawn having the *same* inclination to each other that two *given* lines have. We must therefore first find the inclination of the given lines. To do this we have only to draw on the *given* angle an arc of any convenient radius, with A as centre, such as BC. The length of its *chord* is the distance subtended by the lines forming the angle at the radius AB or AC. If, then, from point *a*, in the line DE, and with a radius equal to AB, we describe an arc *bc*, and from *c* set off a distance on *bc* equal to the chord of the arc BC, then a line drawn through *b* and *a* will make the same angle with DE that AB does with AC in the given angle, which solves the problem.

Problem 8 (Fig. 51).—*To draw a line making a given angle—say 60°—with a given line.*

The solution of this problem involves the relation that the radius of any circle has to the chord of an arc which subtends an angle of 60° in the circle. To solve it, let AB be the given line, and C a point in it at which it is desired to draw a line making an angle of 60° with AB.

MECHANICAL AND ENGINEERING DRAWING   27

From C, as centre, and with a convenient radius, draw the arc DE, cutting AB in E; from E with the same radius cut DE in D, then a line drawn through D and C will make an angle of 60° with the line AB. If the circle were completed with the same radius, it would be found, on stepping the radius round it, that it exactly divides it into *six* equal parts, and as every circle for geometrical purposes (as before explained) is divided into 360°, one-sixth of the circle must contain 60°, or the angle which the two lines in the problem have to make with each other. Knowing this specific relation subsisting between the radius and the chord of an arc of 60° of a circle, we are enabled to lay down any angle with the assistance of a "scale of chords," which will be found on one of the set of drawing-scales previously recommended. To show its use, let us take, for example—

Problem 9 (Fig. 52).—*To draw a line, making an angle of, say, 70°, with a given line at a given point in it.*

Let AB be the given line, and *a* the given point in it. From the zero point, on the *extreme left* of the scale of chords, and with a radius in the compasses equal to the distance from that point to the one marked 60—with the arrow over it—on the scale, draw with *a*, on the line AB as a centre, the arc *bc*, cutting AB in *c*, and from *c* as a centre, with a radius equal to 70° on the scale of chords, cut the arc *bc* in *b*. A line, drawn through *b* and *a* will make, with the given line AB, an angle of 70°; and so with any other angle, always remembering that from *zero to 60* on the scale of chords is the radius with which the first arc in the construction is to be drawn.

CHAPTER V

PLANE GEOMETRICAL FIGURES

13. IT may be noted, before passing on to the construction of the plane geometrical figures which form the surfaces of the plane solids whose projections we shall next show how to obtain, that as the angles most generally chosen for the surfaces of mechanical details are those which contain some multiple of 5°, it is not necessary to use even a scale of chords in laying them down on paper or other material, as most of them can be obtained by simple geometrical construction, which has fewer chances of error than even measuring from a scale.   A few of such angles are 15°, 30°, 45°, 60°, 75°, 120°, 135°, etc., and are thus obtained : For 30°, bisect 60° ; for 15°, bisect 30° ; for 45°, bisect 90° ; for 60°, use radius ; for 75°, add 15° to 60° ; for 120°, mark off radius twice ; for 135°, take 45° from a semi-circle.   With these simple constructions committed to memory, and the use of a scale of chords for any angle not easily obtained otherwise, the student will be able to lay down any angle that may be required.   We may now proceed with the construction of plane figures, taking first—

Problem 10 (Fig. 53).—*To construct an equilateral triangle on a given base.*

(Note : The *base* of any triangle is that side of it on which it stands ; the *vertex*, the point immediately over the base ; and the *altitude* the height of the vertex from the base.)   With the given base AB as a radius, and from A and B as centres, describe arcs cutting each other in C, the vertex, join AC and BC, and the triangle is constructed.   If the altitude only be given as CD (Fig. 54) : Then, as the sum of the angles of any triangle are together equal to two right angles, or 180°, and as the triangle required is equi-angular, the angle at its vertex will be one-third of 180°, or 60°.   To construct it, draw EF, GH through C and D at right angles to CD, and from C, with any convenient radius, describe a semi-circle cutting EF in *a* and *c ;* with the same radius, and from *a* and *c* as centres, cut the semi-circle in *d* and *e*, draw lines through C*d* and C*e*, and produce them to meet GH in *g* and *h*, then *g*C*h* is an equilateral triangle having an altitude CD.

MECHANICAL AND ENGINEERING DRAWING

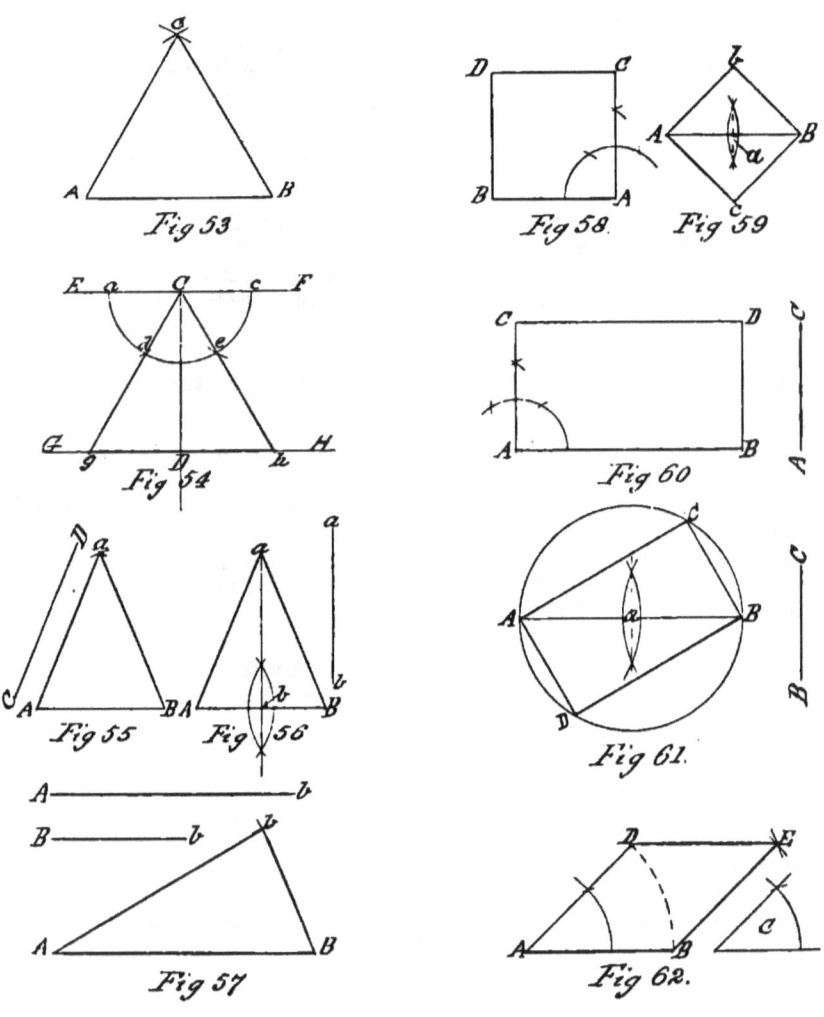

Fig 53

Fig 58

Fig 59

Fig 54

Fig 60

Fig 55

Fig 56

Fig 61

Fig 57

Fig 62.

30   FIRST PRINCIPLES OF

Problem 11 (Fig. 55).—*To construct an isosceles triangle, the base AB and one of the equal sides CD being given.*

With CD as a radius, and from A and B as centres, draw arcs intersecting in *a*, join *a*A and *a*B, and the triangle is constructed. If the base AB and the altitude *ab* are given (Fig. 56): Bisect the base AB in *b*, and at *b* erect a perpendicular and make it equal to *ab*, join *a*A and *a*B, then A*a*B is the required isosceles triangle.

Problem 12 (Fig. 57).—*To construct a scalene triangle, the sides being given.*

Take the longest side AB for the base, and with the shortest as a radius, and from B as a centre, describe an arc ; then with the length of the third side as radius, and from A as centre, cut the arc described from B in *b*, join *b* and A, and *b* and B, then A*b*B is the required triangle.

Problem 13 (Fig. 58).—*To construct a square on a given line AB as a side.*

Erect at A a perpendicular to AB, and from it cut off AC equal to AB; then from C and B as centres, and with AB as radius, draw arcs intersecting at D, join C and D and B and D, and the square is constructed. If the given line be a diagonal and not a side : Bisect the diagonal AB (Fig. 59) in *a*, by a perpendicular *b*, *a*, *c*, and from *a* set off *ab*, *ac*, equal to *a*A, or *a*B, join A*b*, *b*B, B*c*, *c*A, and the square is constructed on the given diagonal AB.

Problem 14 (Fig. 60).—*To construct a rectangle, the length of two adjacent sides being given.*

Let the line AB be one of those sides. At A erect a perpendicular to AB, and cut off from it in C, a length equal to the other given side ; from B as centre, and with a radius equal to AC, draw an arc, and from C as centre, with a radius equal to AB, draw another intersecting the first in D, join CD and DB, and the required rectangle is constructed.

Problem 15 (Fig. 61).—*To construct a rectangle, a diagonal AB and one side BC being given.*

As the *diagonal* of a rectangle divides it into two right-angled triangles, if it is made a diameter, and on it a circle is described, the circle will contain the two right-angled triangles which will form the rectangle sought. Therefore, bisect the given diagonal AB in *a*, and from *a*, with *a*B as radius, describe the circle ABCD; from B as a centre, and with BC as radius, cut the circle in C, and from A, with the same radius, cut it in D, join ACBD, and it is the required rectangle.

MECHANICAL AND ENGINEERING DRAWING . 31

Problem 16 (Fig. 62).—*To construct a rhombus, one of a pair of opposite angles and length of a side being given.*

Let AB be the length of given side, and C the given angle ; at A make the angle BAD equal to angle C, and the side AD equal to AB ; from B and D as centres, with AB as radius, draw arcs intersecting in E ; join EB and ED, and ADEB will be the required rhombus. If a diagonal AB and length of a side AC be given (Fig. 63) : Then, if from A and B as centres, with a radius equal to AC, arcs be struck cutting each other in C and D, and lines be drawn joining A and B to C and D, the figure ACBD will be the required rhombus.

Problem 17 (Fig. 64).—*To construct a rhomboid, the lengths of two adjacent sides and one of a pair of its opposite angles being given.*

Let AB be one (the longest) of the adjacent sides, and E one of the opposite angles. At A make the angle CAB equal to the angle E, and cut off AC equal to the shorter adjacent side. From C, with AB as radius, describe an arc, and from B, with AC as radius, describe another cutting the first in D, join ACDB, and it is the required rhomboid. If a diagonal AB (Fig. 65) and the lengths of two adjacent sides be given : Then, with the length of one of those sides as a radius, and from A and B as centres, describe arcs on *opposite* sides of AB, and from the same centres, with the length of the other adjacent side as radius, describe arcs cutting those first drawn in C and D, join AC, CB,BD,DA, and it will be the required rhomboid.

Problem 18 (Fig. 66).—*To construct a trapezium, the length of its sides and one of its angles being given.*

Let AB be the *base* of the figure or side on which it stands, and C the given angle. At A in AB make DAB equal to the angle C, and let AD equal the length of that side of the figure ; with the length of the opposite side as radius, and from B as centre, describe an arc, and from D as centre, with the length of the fourth side as radius, strike an arc cutting the last in E, join ADEB, and the required trapezium is constructed.

14. In the construction of the preceding plane figures, the lengths of one or more of their sides, with their relation to each other, are previously known or determined by the given problem. In the case of a regular *polygon*, the data generally given are its *kind*, and the length of a side, or a given circle within which it is to be inscribed. The ordinary solution in such cases involves the remembering of certain specific constructions which are liable to be forgotten when most needed. All that is absolutely required to be known for the construction of *any* regular polygon, is the *relative position* of any two of its adjacent sides, and in certain cases the *length* of one of them.

The relative position, or, in other words, the *angles* made by any two adjacent sides of a regular polygon, are easily determined. The

32   FIRST PRINCIPLES OF

exterior angle, or that formed by one side with the other produced, is always equal to 360° divided by the *number of the sides* of the polygon, and the *interior* angle, or that formed by the meeting of the two adjacent sides, is 180° minus the exterior angle. The angle at the centre (or central angle) of a regular polygon is equal to the exterior angle. With these simple facts committed to memory, the student or apprentice can, with a scale of chords—now generally found on all pocket rules,—lay down at once on his work any regular polygon having either an odd or an even number of sides. To apply these facts we will take—

Problem 19 (Fig. 67).—*To construct a regular pentagon with a given length of side.*

Here 360° ÷ 5 equals 72°, the exterior angle; and 180° ± 72° = 108°, the interior. Let AB be the given side, produce it (say to the left) at A, draw the line AC, making an angle of 72° with AB produced, and of a length equal to AB; bisect AB and AC by perpendiculars intersecting in S, then S is the centre of the circumscribing circle. Describe it, and from C, with AB as a distance, set off on it the points D, E, join CD, DE, EB, and ACDEB is the required pentagon.
If the pentagon has to be *inscribed* in a given circle, then from its centre—which will be the centre of the pentagon—draw any radius as SA (Fig. 67) at S, draw a line making with SA an angle of 72°, and cutting the circle in B, join A and B, then AB is one side of the required pentagon; set off the distance AB from A or B round the circle, and it will give points C, D, E; join ACDEB, and the pentagon is constructed in the given circle.

Problem 20 (Fig. 68).—*To construct a regular hexagon with a given length of side.*

Here 360° ÷ 6 equals 60°, and 180° – 60° = 120°. Let AB be the given side, produce it, and draw AC, making with AB produced an angle of 60°; make AC equal to AB, bisect them by perpendiculars intersecting in S, which is the centre of the circumscribing circle; describe it, and set off the distance AB round it from C, in points D, E, F, join CD, DE, EF, FB, and the required hexagon is constructed. If a hexagon has to be *inscribed* in a given circle, the central angle will be 60°; this angle laid down with the centre of the circle as the angular point will give A, B (Fig. 68), points in the circle, and the line joining them will be a side of the hexagon; step this length round the circle in points C, D, E, F, join AC, CD, etc., and the required hexagon is inscribed in the given circle. As the side of a hexagon is the chord of an arc of 60°, and is equal to the radius of the circumscribing circle, that radius set off round the circle will divide it into six equal parts, and if the points of division be joined by right lines they would form the inscribed hexagon as before.

MECHANICAL AND ENGINEERING DRAWING   33

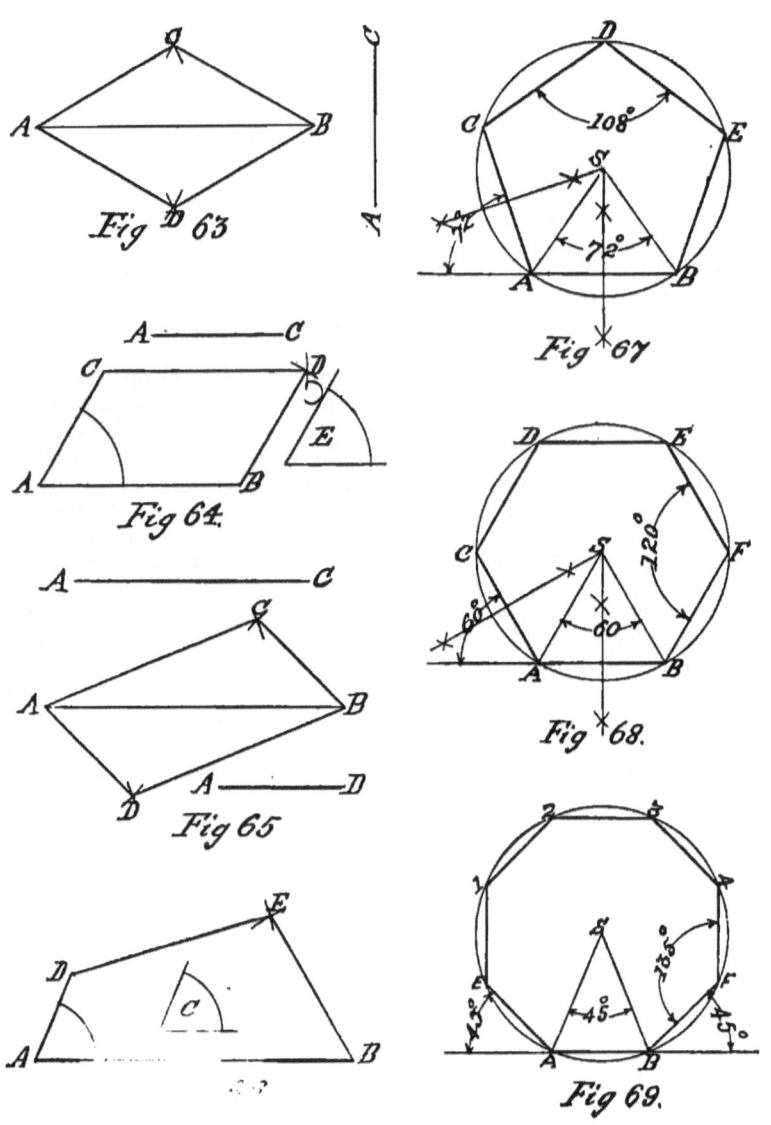

Fig 63

Fig 64.

Fig 65

Fig 67

Fig 68.

Fig 69.

D

Problem 21 (Fig. 69).—*To construct a regular octagon, with a given length of side.*

Here $360° \div 8 = 45°$; and $180° - 45° = 135°$. Let AB be the given side. Produce it in both directions, and at A and B draw lines AE, BF, of the same length as AB, and making with AB produced angles of 45°; bisect the angles formed at A and B, and their intersection at S will be the centre of the circumscribing circle. With SA or SB as radius, describe this circle, step AB round it from E to F in the points 1, 2, 3, 4; join E 1, 2, 3, 4 F, and the required octagon is constructed. To inscribe an octagon in a given circle: Draw two radii (Fig. 69) at an angle of 45° to each other, and they will cut the circle in points A and B; join AB, and it will be a side of the octagon. Its length stept round the circle will give the same points as in the previous construction; join them, and an octagon will be inscribed in the given circle.

· **15.** The same principle of construction as used in the last three problems is applicable to any regular polygon, whatever may be the number of its sides; but in practice it is preferable to subdivide the sides of those we have given—if the division will give the required number of sides—than to lay down an independent construction, the chances of *not* obtaining the exact length of the side of the polygon required increasing as the number of sides increase. On paper, and with the assistance of tee- and set-squares, many of the figures already given can, of course, be easily and quickly constructed; but, as before observed, the ability to draw them *without* such aids is absolutely essential, when we consider the calls often made upon the workman for the practical application of such knowledge.

As figures, or solids, having more than eight sides or plane surfaces are seldom met with in mechanical construction, and as those we have given include all that form the surfaces of the plane solids intended to be used as objects for projection, we shall now proceed to show how their projections are obtained.

CHAPTER VI

ORTHOGRAPHIC PROJECTION

16. A CAREFUL study of the preceding chapters, and the solution of the problems contained in the two last, will have prepared the student for entering upon that more important part of our subject— viz., "Orthographic Projection," or that special kind of delineation which, when applied to the representation of mechanical subjects, enables the engineer or machinist to determine at sight the actual dimensions and arrangement of any part of an engine or machine. As, however, a *part* of a piece of mechanism is but a compound of simple forms made up of what are known as plane solids and solids of revolution— alone or combined—it is at once manifest that to be able to draw any part of a machine, the would-be draughtsman must first master the delineation of its component parts, and as these resolve themselves into solids, with either plane or curved surfaces, having straight or curved lines for their boundaries, the question of their ultimate accurate representation as a whole becomes one of the *correct projection* in the first stage of the study, of the lines bounding the *surfaces* of solids ; and as straight lines and flat surfaces are more easy of projection than curved ones, we commence this part of the subject by an illustration of its principles in the projection of points, straight lines, and the simple figures which form the surfaces of those plane solids used in giving shape to machine and engine details.

By a reference to the latter part of Chapter II., it will be noted that to obtain the views of an object required for the purposes of manufacture its. projections are determined on two planes, at right angles to each other—that is, their relative positions are as shown in Fig. 13 ; that lettered **VP** being a plane assumed to be vertical, and the other **HP** a horizontal plane perpendicular to VP. These two planes are called the *vertical* and *horizontal* "planes of projection," and will throughout the exposition of the subject of "projection" be denoted by the letters VP and HP.

17. The student should be particular to note the precise difference of meaning existing between a "vertical" line or plane and one that is "perpendicular." One line or plane may be perpendicular to another line or plane, and yet neither of them be vertical. A *vertical* line is a

36   FIRST PRINCIPLES OF

plumb line, or the position a weighted line assumes when freely
suspended. A *horizontal* line is one which is parallel to the horizon,
and, therefore, perpendicular to a vertical one. A "vertical plane,"
then, is one with which a plumb line will coincide, and similarly a
"horizontal plane" is one parallel to the earth's surface *taken as a
plane*, and is at right angles to the vertical.

A *plane*, strictly defined, is nothing more than a perfectly flat "surface," without any reference to *substance ;* but as it cannot be dealt
with for explanatory purposes without being assumed to be material
and inflexible, it will, when spoken of, or used for that purpose in this
work, be considered as having such a thickness as would be represented by a line. Assuming this, the *edge* view of a plane will, under
any circumstance of position, be a perfectly straight line. If, then, two
planes *intersect* or meet each other at an angle, as the "planes of projection" we are about to deal with do, their meeting will be in a line,
which forms a boundary or dividing line between them, and is called

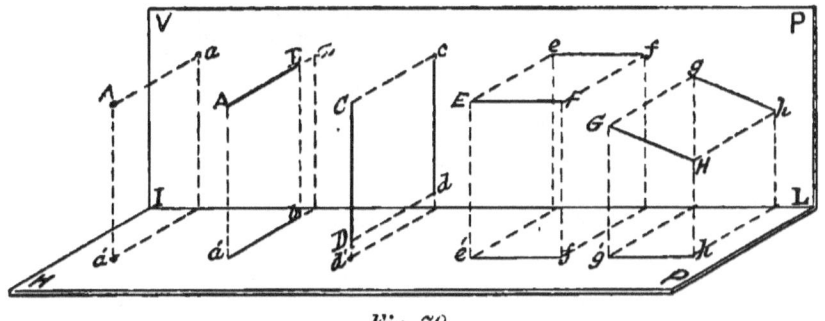

Fig. 70

the "intersecting line" of the planes. This line will throughout the
subject have IL for its distinguishing letters.

Knowing, then, the true relative position of the "planes of projection" on which we wish to obtain the representations of an object,
we will first proceed to find the projections of a "straight line" in
different positions with respect to those planes. Let its position at
first be perpendicular to the VP.

Here, as the thing to be projected is a "line" having ends or points,
before we can obtain its projections we must first know how to find
those of a "point." Let, then, A on the left in the diagram Fig. 70 be
a point in space, such as a small bead invisibly suspended, and let it
be required to find its vertical and horizontal projections—that is, its
projections on the VP and HP.

To obtain these, we have to find the points in the VP and HP
where a visual ray or projector perpendicular to each of the planes, and
drawn through A, would penetrate them. This, it will be seen in the
diagram, is in *a* in the VP, and *a'* in the HP, and therefore *they* are

MECHANICAL AND ENGINEERING DRAWING   37

the required projections, *a* being an *elevation* or vertical projection of
A, and *a'* its *plan* or horizontal projection. If it were required to find
from its projections the *position* of the original point A with respect to
the VP and HP, then perpendiculars to those planes let fall from its
projections *a* and *a'* would intersect in A, giving it as the position of
the original point.

Knowing how to obtain the projections of a point, we shall now be
able to find the projections of a straight line.

1st. Let the line AB (Fig. 70) be perpendicular to the VP.

Here AB being perpendicular to the VP, will be parallel to the
HP; therefore, from its position with respect to the VP, its projection on that plane will become a point *a*, as the eye being directly
opposite the end of it, the visual ray or projector proceeding from the
eye will travel along the line itself, coinciding with it, and penetrate
the VP in *a*, then *a* is the "elevation" of the line AB. To find its
"plan" or projection on the HP, let fall projectors perpendicular to
the HP from both ends of AB, and the points *a' b*, where these projectors penetrate the HP, will be projections of the ends A and B of
the line AB, and if *a' b* be joined, then *a'b* will be the plan or horizontal projection of the original line.

2nd. Let CD (Fig. 70) be the given line, and let it be perpendicular to the HP, and its projections required.

In this case CD is parallel to the VP, and its projection on that
plane will be obtained by letting fall from C, D its ends, projectors to
the VP, and the points *c d*, where these fall on that plane, will be the
vertical projections of C and D in it; then, if *c* and *d* be joined, *c d*
is the elevation of the line CD. As the given line is perpendicular
to the HP, its plan will be a point obtained by producing a visual ray
passing through and coinciding with CD itself, until it penetrates the
HP in *d'*.

3rd. Let EF be the given line (Fig. 70), and let it be parallel to both the VP and the HP, and its projections required.

Here EF being parallel to both planes, by letting fall projectors
from E and F to the VP, we obtain points *e* and *f*, and to the HP
points *e'* and *f'*, then *ef* and *e'f'* being joined will be the required
projections. It will be noted here that the projections of the original
line EF are two lines of the same length as their original. This is
owing to the relative positions of the original line, and the planes on
which its projections were required. Had the line been in any other
position with respect to those planes, a different result would have been
obtained, as will be seen by the following problem.

38   FIRST PRINCIPLES OF

*4th. Let GII be the given line, and let it be parallel to the VP, but inclined
to the IIP, and its projections required.*

Here GH being parallel to the VP, its vertical projection or elevation is found by letting fall projectors from G and H to the VP, giving
points *g* and *h*, which, when joined, will be a line of the same
length as GH; but its horizontal projection, obtained by letting fall
projectors from the same points G and H on to the HP, giving *g'h'*, will
be found to be projected into *g'h'*, a much shorter line than its original.

The diagram Fig. 70, the student must note, is drawn in what is
known as "quasi-perspective," and is adopted as a simple and ready
means of showing the two planes of projection in their relative positions,
and the positions of the lines given in the foregoing problems in relation

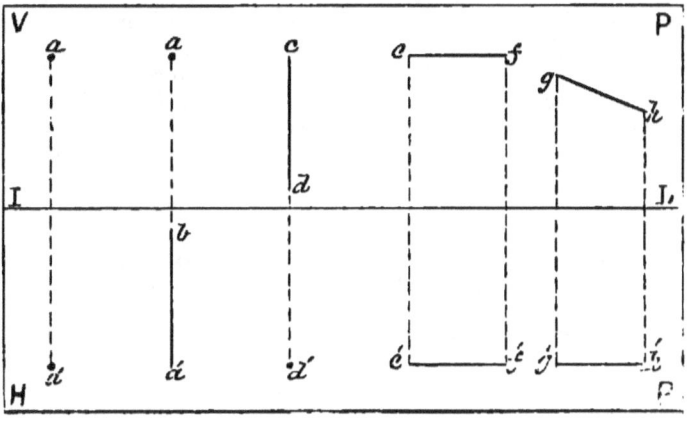

Fig. 71

to those planes. It is in no sense an orthographic projection diagram,
although used to explain the application of the principles of that kind
of projection.

17. To convert the actual relative positions of the two planes of
projection, as shown in the diagram Fig. 70, into the positions they
occupy on the sheet of drawing-paper when laid on his board, the
student has to suppose the "upper" plane, or that we have named the
VP, turned backwards on the IL (intersecting line) as a hinge, until it
is on the same level with the "lower" one or the HP, the two planes
thus becoming one flat surface, as in Fig. 71, with the IL dividing them,
and the plans and elevations of the lines in the problems shown on them
as obtained by projection.

Assuming that the student has found no difficulty in understanding
the explanations already given of the way in which the projections of a
line when it is in either of the suggested positions, with respect to the

MECHANICAL AND ENGINEERING DRAWING   39

planes of projection, are obtained, there are yet two other positions that a line may occupy with respect to those planes, whose projections we must know how to find before we can proceed with the projection of plane figures.   One of those positions is that of a line inclined to *both* the VP and the HP.   We have shown, in Figs. 70 and 71, that if a line be parallel to one plane and inclined to the other, its projection on the plane to which it is parallel will be a line equal in length to the original, and on the one to which it is inclined its projected length will depend upon the angle the given line makes with its plane of projection. This will be made still clearer by the demonstration of the problem where—

5th. A line AB is inclined to both the VP and HP (Fig. 72), and its projections are required.

Let the given line at first be parallel to the VP, and perpendicular to the HP.   In this position its projection on the HP will be a point, as *a*, and on the VP a line AB, at right angles to the IL.   While keeping AB parallel to the VP, conceive it to swing round to the right on A as a joint, until it makes any desired angle with the IL ; or say until B has moved into the position *b*, its elevation *b*A in this position is a line inclined to the IL, of the same length as AB, but its plan, obtained by letting fall from *b* a projector perpendicular to the HP, or IL, in *c*, gives *ac* as its projection on the HP, or a line less than half the length of its original.   It is evident from this that the *projected length* of a line is entirely dependent upon its angle with the plane of its projection, for if the motion of the line AB in this case were continued until it coincided with the IL, its projected length *ad*, and its original length AB, would then become equal.

But so far the given line is only inclined, as at A*b*, to *one* of the planes of projection, the HP ; for although we have moved it from its assumed first position—that is, perpendicular to the HP—to that of making an angle *b*AD with it, it is still parallel to the VP.   Let it also be inclined to that plane, say 45°.   For distinctness, let C be a new position of A on the IL ; at C draw C*t*, at an angle of 45° with the IL and equal to *ac*, or the projected length of A*b* in the HP ; then C*t* will be a plan of A*b* when at 45° to the VP, and at the angle *b*AD with the HP.   To obtain its elevation, draw from *t* a projector perpendicular to the IL, and from *b* another parallel to it, to cut the one from *t* in *p*, join C and *p*, and the line C*p* will be the elevation of the original line AB, inclined to both planes of projection.   Here it will be noticed that the original line AB, in addition to its having been moved on A as a joint from B to *b*, has also, while making the angle *b*A*d* with the HP, been swung round on A through 45°.

Now to make this matter of the projection of inclined lines still more clear, as much depends on the student having a thorough grasp of this first part of the subject.   We will assume that the two projections, C*p* in the VP, and C*t* in the HP, are *given*, and it is required to find the *real length* of the line of which they are the projections.

Here the line C*t* is the plan or horizontal projection of the line C*p*,

40   FIRST PRINCIPLES OF

the latter being a line having *one* of its ends C, *in* the HP, and the other
end *p* a given distance *above* that plane.   C*p* is also the projected length
of the hypothenuse (or longest side) of a right-angled triangle, having
C*t* for its base, and a line equal to the vertical height of *p* from the HP
for its perpendicular.   With these two sides given, we can find the third
side, or the actual line of which C*p* is the projection.   Therefore, at *t* in
the line C*t*, and perpendicular to it, draw a line indefinitely, and from it
cut off in *h*, a length *th* equal to the height that *p* in the line C*p* is above
the HP or IL, join C and *h ;* then C*h* is the real length of the original
line, of which C*p* and C*t* are its projections.   This is self-evident, for if
the right-angled triangle C*th*, which may be assumed to be lying on the
HP, with its base line coinciding with C*t*, be raised to a vertical position,
moving on C*t* as a hinge, its base and hypothenuse will then be coincident
with C*t*, and its third side *ht* is a vertical line perpendicular to the HP
represented by the point *t*.

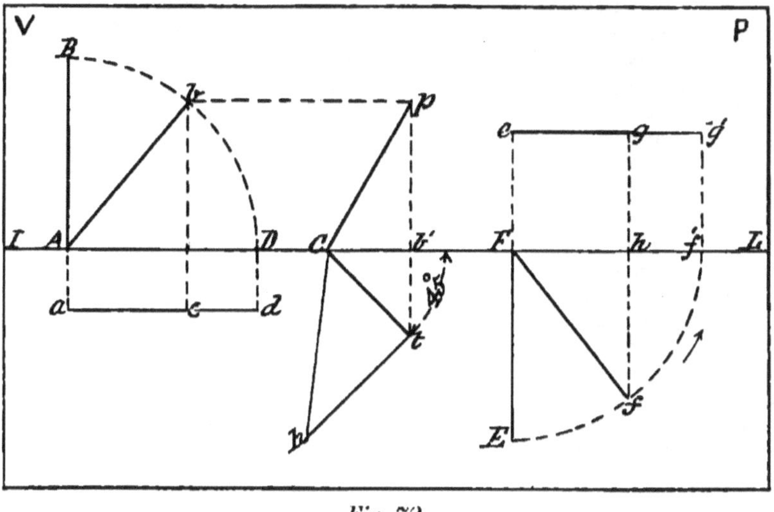

Fig. 72.

The other position a line may have, with respect to the planes of its
projection, is that of being parallel to the HP, but making an angle with
the VP.   Putting this in the form of a problem, we will say—

*6th. Let a given line be parallel to the HP, but inclined to the VP, and its
projections required.*

In this case, let the given line at first be perpendicular to the VP ;
its elevation when in that position in the VP will be a point as *e*, and
its plan a line EF at right angles to the IL.   But as EF is perpendicular
to the VP it is parallel to the HP.   While keeping it so, let it be conceived to swing on its end F as a joint in its direction of the arrow,
until it makes any desired angle with the VP or IL, or until, say, E

MECHANICAL AND ENGINEERING DRAWING   41

has moved into the position f; its elevation in that position is found by drawing a projector through f, perpendicular to the IL, and a line through e parallel to it to cut the projector from f in g, then the line eg is the vertical projection of EF when making the angle LFf with the VP. Here it is again seen that the projected length of a line, although parallel to one of its planes of projection, is determined on the other plane by the amount of its inclination to that plane ; for had the given line EF in this case been moved through any greater or less angle than the one assumed in the diagram, its projected length in the VP would have been greater or less than eg, directly in proportion to its altered position with respect to the VP. If EF had been swung so far round on F until it had coincided with Ff, then its projection in the VP would be eg', or a line equal in length to the given line EF.

18. In the foregoing problems in this chapter, we have given all the positions which it is possible for a line to occupy with reference to its planes of projection, and we could at once proceed to the projection of *plane figures*, were it not necessary at this stage that the student should thoroughly understand the true significance of the line lettered IL in previous and all future diagrams throughout this work.

This line, we have already shown, is the line of intersection of the two "planes of projection"; but it is much more than this. The VP and HP, being for all the future purposes of the student draughtsman represented by the *one* flat surface of his sheet of paper, with the IL either shown on it or assumed to be there, dividing its surface into two planes, it becomes—when *shown in* on a drawing or explanatory diagram —at one and the same time the representative, not only of the IL, but of the VP and HP as well, for it is a plan of the VP and an elevation of the HP, and as these it is a *datum* line from which heights *above* the HP, or distances *from* the VP, may be measured or set off.

These facts, it will be seen, are verified by a reference to Fig. 72. Here the IL is for all the figures, a *plan* of the VP, showing the line Ab by its plan to be parallel to the VP, and in front of it, at a distance equal to Aa. Similarly C, the lower end of the line Cp, is shown *touching* the VP, while its upper end p, projected into t in the HP, stands *out* from the VP a distance equal to tt', and *above* the HP a height equal to $t'p$. Then again Fe is the height of the line EF above the HP, the end F touching the VP in e, and the end E being a distance equal to EF from the VP when in the first position, and a distance equal to fh when in the second. In the two last cases, it will be seen that the IL is a plan of the VP and an elevation of the HP. With the foregoing explanation of the projection of lines thoroughly digested, the student should have no difficulty in finding the projection of plane figures, to which we now proceed.

CHAPTER VII

PROJECTION OF PLANE FIGURES

19. Knowing how to find the projections of a line having any given position with respect to the planes of its projection, we can now proceed to the projection of those straight-sided plane figures which form the surfaces of the solids used in giving shape to machine details. As the same principles apply to the projection of all plane figures, whether their sides are few or many, it is only necessary that their application should here be shown in the case of one of each class of figure chosen. Commencing with that figure having the least number of sides—the triangle —we shall give as additional subjects for projection the square, rectangle, pentagon, and hexagon, or those which usually form the sides and ends of the plane solids we have before referred to. Our first problem in this subject is—

Problem 22 (Fig. 73).—*The line a c is the plan of an equilateral triangle, with its base resting on the HP, and parallel to the VP; it is required to find its elevation or vertical projection.*

Here, as the base of the triangle is parallel to the VP and its plan is represented by a *line*, the triangle itself is parallel to the VP, and therefore perpendicular to the HP. To find its elevation, through *a* and *c* draw projectors perpendicular to the IL, cutting it in A and C; through A and C with the 60° set-square draw lines intersecting in B, join A, C, and the figure ABC is the required vertical projection of the triangle of which the line *a c* is the plan. The projection, it will be seen, is an *equilateral* triangle, for all the sides of the triangle being by the conditions of the problem parallel to the VP, they are projected on that plane into lines of the same length.

Problem 23 (Fig. 74).—*To find the elevation of the triangle obtained in the previous problem when it is inclined 45° to the VP, its plan being the line a c, as before.*

Here the IL may be considered as a plan of the VP. If we draw a line (with the 45° set-square) of a length equal to *a c*, at an angle of

42

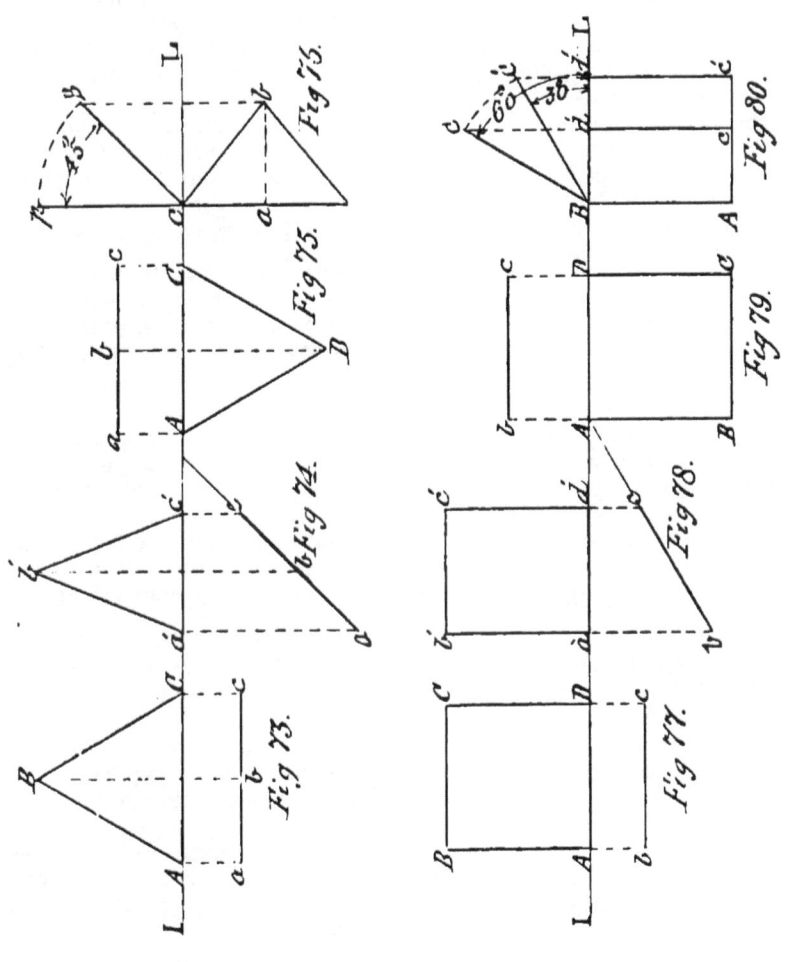

44   FIRST PRINCIPLES OF

45° with the IL, that line will be a plan of the triangle ABC when at that angle with the VP. Bisect *ac* in *b*, through *b* draw a projector perpendicular to the IL, and from B, in Fig. 73, another parallel to it, cutting the one drawn from *b* in *b'*. Then, from *a* and *c* draw projectors to the IL, cutting it in *a'* and *c'*, join *a'b'*, *a'c'*, *c'b'*; the figure *a'b'c'* is the required elevation of the given triangle inclined 45° to the VP. In this case it will be noticed that the elevation obtained is an *isosceles* triangle, resulting from the altered position of its original with respect to the VP, the plane of its projection. The line *ac* is bisected in *b* to find the plan of the vertex B of the triangle ABC, Fig. 73; and the vertical projector through this bisection *b* determines, by its intersection with a parallel one through B, the elevation *b'* of the vertex in its new position.

Problem 24 (Fig. 75).—*The line a c is the elevation of an equilateral triangle having its base touching the VP and parallel to the HP; to find its plan or horizontal projection.*

The position of the original figure, of which the line *a c* is the elevation, is in this case the converse of that in Fig. 73. Here *a c* being a line in the VP parallel to the IL, the triangle whose projection it is must be parallel to the HP. To obtain its plan let fall from *a* and *c*, its ends, projectors perpendicular to the IL, cutting it in A and C, and through AC with the 60° set-square draw lines intersecting in B, then the figure ABC is the required plan of the triangle of which *a c* is the elevation.

Problem 25 (Fig. 76).—*Let the triangle obtained in the last problem be inclined to the HP at 45°, its base resting on that plane at right angles to the VP, and one angular point touching the VP; to find its projections in that position.*

Assume the position of the triangle at first to be perpendicular to both the VP and HP; its elevation will then be a line perpendicular to the IL, equal to the altitude of the triangle, as C*p*; and its plan, a line AC, also perpendicular to the IL and equal to the base of the triangle. If, then, the triangle be moved on AC as a hinge (to the right) through 45°, its elevation at that angle is found by drawing—with the 45° set-square—through C a line CB equal to C*p*. To find the plan, bisect AC in *a*, and through *a* draw a line parallel to the IL: let fall from B a projector perpendicular to the IL to cut the line drawn through *a* in *b*, join C*b* and A*b*, and the figure CA*b* is the projection of the triangle ABC (Fig. 75) when inclined at 45° to the HP.

Problem 26 (Fig. 77).—*Given a straight line b c parallel to the IL as the plan of a square resting with one of its sides on the HP; to find its elevation.*

The plan of the square being a line parallel to the IL, the square itself will be parallel to the VP and perpendicular to the HP; there-

MECHANICAL AND ENGINEERING DRAWING   45

fore, from b and c draw projectors perpendicular to the IL, cutting it in A and D; set off on one of them a distance AB equal to bc in the plan, and through B draw BC parallel to the IL; join AD, and the figure ABCD is the required elevation of the square whose plan is the line $b\ c$.

Problem 27 (Fig. 78).—*To find the elevation of the square obtained in the last problem when it is inclined at 30° to the VP, its plan being a line $b\ c$, as before.*

Draw in the HP (with the 60° set-square) a line $b\ c$—the plan of the square—making an angle of 30° with the IL; and from b and c draw projectors cutting the IL in a' and d', make $a'b'$ equal to bc, and through b' draw $b'c'$ parallel to $a'd'$, then the figure $a'b'c'd'$ is the elevation required.

Problem 28 (Fig. 79).—*A square with one of its sides touching the VP is represented in elevation by a line $b\ c$ parallel to the IL; to find its plan.*

From b and c let fall projectors perpendicular to the IL; make AB equal to bc, and through B draw BC parallel to the IL, join ABCD, and it is the required plan of the square.

Problem 29 (Fig. 80).—*The square ABCD obtained in the last problem is inclined at 60° and 30° to the HP, with one of its sides touching the VP and an adjacent side the HP; to find its plan and elevation in those positions.*

At B in the IL, draw BC, BC' (both equal to AB) with the 60° set-square, making angles of 60° and 30° with the IL; then BC, BC' are the required elevations of the square at those angles. From Bdd', draw BA, dc, $d'c'$, perpendicular to the IL, and each equal to BC; and Acc' parallel to it; then BAcd; BA$c'd'$ are the plans of the given square when inclined at 60° and 30° to the HP. The projections of the rectangle are not given, as the method of obtaining them is in all respects the same as that for the square, but allowing for the difference of length of adjacent sides.

20. The method of obtaining the projections of a pentagon and hexagon in different positions with respect to the VP and HP is fully shown in Figs. 81—88, and will not need explanation further than to say that the elevations in Figs. 81 and 85—shown with reference letters in capitals—must be drawn *first*, according to the construction rules given in Problems 19 and 20, before their plans can be found; in all other respects the procedure is the same as in the previous figures.

To make himself thoroughly conversant with the application of the principles of projection to the delineation of plane figures, which it is most important he should be before passing on to the projection of solids, the student should draw all the foregoing plane figures at least twice, making them three times the size here given.

46 MECHANICAL AND ENGINEERING DRAWING

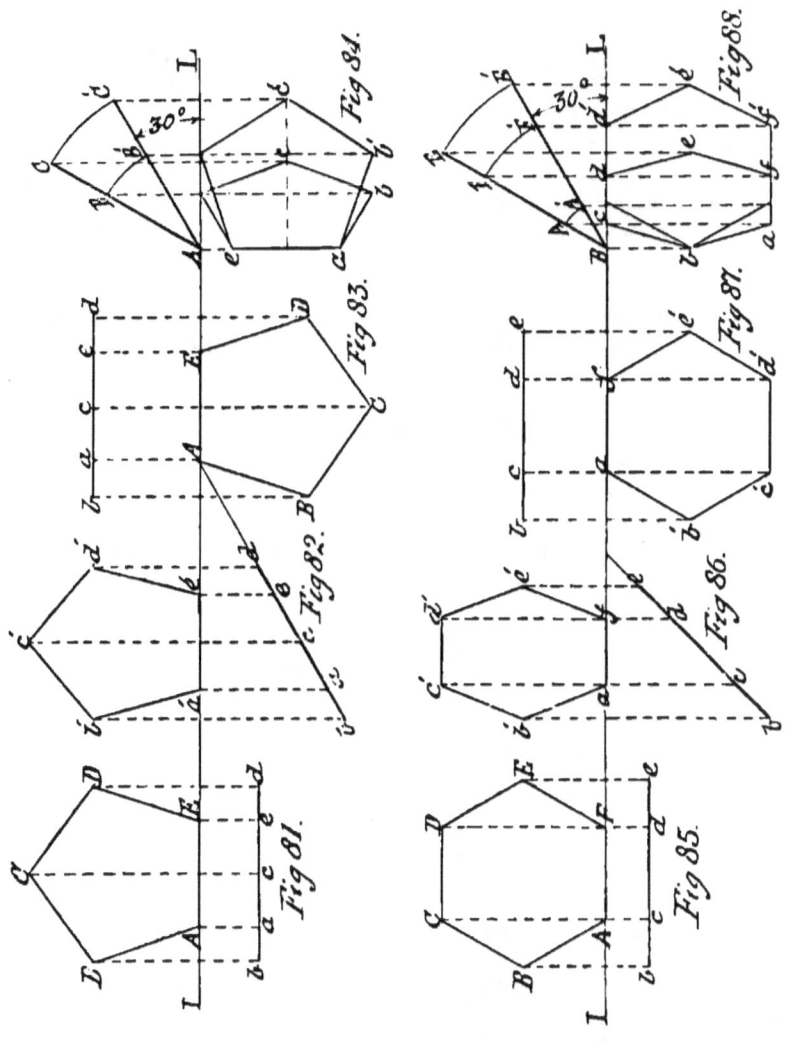

CHAPTER VIII

THE PROJECTION OF SOLIDS

21. ASSUMING that the student has followed the advice given in the last paragraph of Chapter VII., and thoroughly mastered the elementary principles of projection which we have expounded in it and Chapter VI., we can now proceed to apply those principles to the delineation of the simple geometrical solids of which engine and machine details are invariably made up. These solids are of two kinds—viz., plane, and circular or curved. The first-named have all their surfaces plane figures, the projections of which the student already knows how to obtain, and the second includes all solids whose bounding surfaces are all curved, or plane and curved combined. What are known as the simple solids are the cube, the prism, the pyramid, the cylinder, the cone, and the sphere, the first three being *plane* solids, and the others *circular* or curved solids.

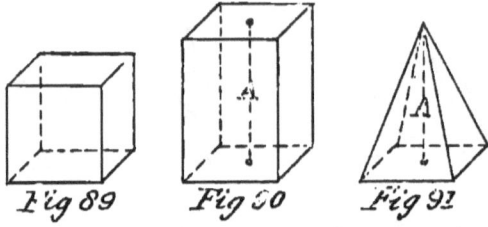

A *cube* (Fig. 89) is a solid having six equal sides or faces, all of them squares.

A *prism* (Fig. 90) has two ends or bases parallel to each other, each being equal and similar figures; its sides are rectangles.

A *pyramid* (Fig. 91) has one base, its sides being triangles, with their vertices meeting in one point *a*, called the *apex* of the pyramid. (As it is advisable for the student to confine himself for the present to the study of the projection of *plane* solids, we defer any consideration of the *circular* ones until after the projection of *curved lines* in any position is understood by him.)

48   FIRST PRINCIPLES OF

Regular prisms and pyramids have "regular" figures for their bases.

The *axis* of a prism, or pyramid, is an imaginary line joining the centres of the bases of the former, and the centre of the base and the apex of the latter, as the dotted lines AA in Figs. 90 and 91.

A *right* prism, or pyramid, has its axis perpendicular to its base, as Figs. 90 and 91. If its axis is inclined to its base, the prism or pyramid is *oblique*, as in Figs. 92 and 93.

A *truncated* pyramid, or prism, is the part of the solid left when its *upper* part is cut away by a plane, and is called a *frustum*. The cutting plane may be either parallel or inclined to the base of a pyramid, but only inclined to the bases of a prism. In Figs. 94 and 95, A and B are frustums.

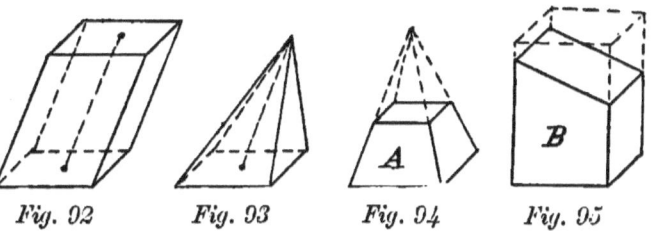

Fig. 92   Fig. 93   Fig. 94   Fig. 95

As all the sides of a prism are parallel to its axis, the edges of the sides connecting its bases are perpendicular to the bases in a "right" prism, and inclined to them in an "oblique" one.

In a right pyramid all its sides are isosceles triangles, and its axis is perpendicular to its base. If the base is a regular figure and the axis perpendicular to it, the sides of the pyramid will all be equal and similar isosceles triangles, but if the axis be inclined to the base, then the sides become unequal triangles and the pyramid an oblique one.

Both prisms and pyramids are named according to the figures of their bases. If the base is a triangle, square, pentagon, hexagon, or octagon, then the solid becomes a triangular, square, pentagonal, hexagonal, or octagonal pyramid or prism, as the case may be.

There are other plane solids, such as the tetrahedron, octahedron, etc., etc., but such forms are seldom adopted by the engineer or machinist in his constructions. Their special features may, however, be studied to advantage by the student draughtsman in his spare time.

22. The working out of the problems in Chapter VII. will have shown how the projections of the figures chosen are obtained, and as they form the *surfaces of the solids* used in giving shape to machine details, our next step is to show how to obtain the projections of such solids in *any* given position. Now, as a sketch is often a much more satisfactory means of explaining a method of procedure than many words, we give in Fig. 96 a perspective view of the planes of projection, and the construction or working lines, showing how the plans and elevations of a few simple objects (all bounded by plane surfaces) are

MECHANICAL AND ENGINEERING DRAWING   49

obtained, which will, we think, materially assist the student in his study
of the application of the principles involved. The objects chosen for
the illustration of these principles are simple prismatic solids, or a combination of such, and only require for the comprehension of the method
of their projection, such a knowledge of principles as the student—from
what has gone before—should have now acquired.

The diagram is so plain that it hardly requires explanation; but as
it is important that the procedure in obtaining the projections should
be thoroughly understood, we will at the risk of repetition endeavour to
make it, if possible, still more intelligible. The constructions to the
left of the diagram are a repetition of the three first problems in Fig.
70, but show more fully, by means of the arrows, the direction of the
projectors or *visual rays* with respect to the VP and HP. After the
very full explanation given in Chapter VI. of the way of obtaining the
projections of a *straight* line in any position, nothing more need be said
in reference to it here than that the *positions* of the lines given in those
problems are the positions of the edges—which are all straight lines—
of the prismatic solids given in the diagram as the subjects for
projection.

The first solid, whose projections in the VP and HP of the
diagram are figured 1, 1, is that of a right prism (of any material
substance—say wood); its bases or ends are rectangles, square with
its sides, and its position with respect to the planes of projection is
such, that its sides are perpendicular and its ends parallel to the VP,
its upper and under sides are parallel to the HP, while its other
two sides are perpendicular to it; its ends being parallel to the
VP are also perpendicular to the HP. As the sides and ends of
the solid are plane rectangular surfaces, and in known position with
respect to the VP and HP, their projections on those planes are
obtained in the same way as those of any plane figures of the same
form, and in the same position.

23. At this stage in our subject it will have become apparent
to the would-be draughtsman that he must either be possessed of a
perfect knowledge of the *forms* of the solids he is attempting to
delineate, or have models of them to guide him in his delineations;
in other words, he must have either a true conception or a possession
of the object he wishes to draw, for it is evident from the diagram
we are proceeding to explain, that without a model of the object to
be delineated, or its conceived counterpart, neither plan nor elevation
of it could be obtained. This is one great reason why an earnest
student of mechanical drawing should at this stage in his study
possess himself of a convenient set of models of the solids enumerated
in a previous paragraph, for no greater mistake can be made in the
study of the "projection of solids" than that of making a servile
copy of *any* diagram or drawing to be found in text-books on this
subject. With this slight digression—which has been necessary in
the interest of students of projection—we proceed with the explanation
of our diagram.

The form and dimensions of the solid having been predetermined,
and its position with respect to the VP and HP known, its

E

50   FIRST PRINCIPLES OF

projections—plan and elevation—are obtained as shown. Projectors, or *visual rays* (shown by the dotted lines in the digram), are let fall perpendicularly, as shown by the arrows, from the principal points in the object—which are the ends of the lines or edges bounding it—upon the VP and HP. The points on these planes where the projectors fall are each the plan, or elevation, as the case may be, of the original line end, or edge end, and these points or projections being joined by what their originals are connected with —viz., straight lines—give the plan and elevation required. The other objects in the diagram, whose projections are numbered 2, 2 ; 3, 3 ; 4, 4, are all prismatic in form, and represent a carpenter's pencil, a hollow wooden tube, and the lower end of a square post. Their projections are all obtained in the same way as those just explained, numbered 1, 1, so there is no necessity for their demonstration. The actual rendering of the true projections of the four objects chosen for illustrating the subject of this chapter will now be given with the sheet of drawing-paper as one plane surface.

To obtain the correct projection of the four objects shown in the diagram Fig. 96, we must assume that we have their exact models before us, and are able to draw in the VP on the sheet of paper a full-size view of each of their *ends*—such, in fact, as are shown at A, B, C, D, Fig. 97. Each of these views is an end elevation, or vertical projection, of the "original object" or model, and will be found to agree with those shown in the diagram Fig. 96, on the "Vertical Plane of Projection," and numbered 1, 2, 3, 4. The height of these views above the IL on the sheet of paper is immaterial ; but whatever it is, it should be understood that the objects represented by A, B, C, D are the same height above the HP as these end views of them are above the IL of the VP and HP. Having these end views given, and knowing by measurement the *length* of the models, we can find their "plans" or projections in the HP.

Now A, B, C, D are the front end views of objects in front of the VP, and as the objects are a given length, their back ends must either be assumed to touch the VP, or to be a given distance from it. Assume them to be as in the diagram Fig. 96—viz., a certain distance *from* the VP ; in this case the ends will all be in a straight line that distance from the IL, for, as before shown, the IL is a "plan" of the VP. Taking, then, the IL as a datum line, set off from it the distance it is intended the back ends of the objects shall be from the VP, and through the point set off draw with the tee-square a faint line parallel to the IL, as *b*, *e ;* then as the objects are assumed to be of the same length, and their front ends (or those nearest the eye) are parallel to their back ends, set off from the line *b*, *e*, a distance equal to the length of the objects, and at this distance draw another faint line *f*, *e*, parallel to *b*, *e ;* then the plans of the front ends of A, B, C, D will, in the HP, be *in* the line *f*, *e*, and of their back ends in the line *b*, *e*. From the points 1, 2 of A, in the VP, let fall projectors perpendicular to the IL into the HP, cutting the faint lines *b*, *e* and *f*, *e* in points 1', 1 ; 2', 2 ; join these by straight lines as shown, and the plan, or horizontal projection of the object, whose

MECHANICAL AND ENGINEERING DRAWING

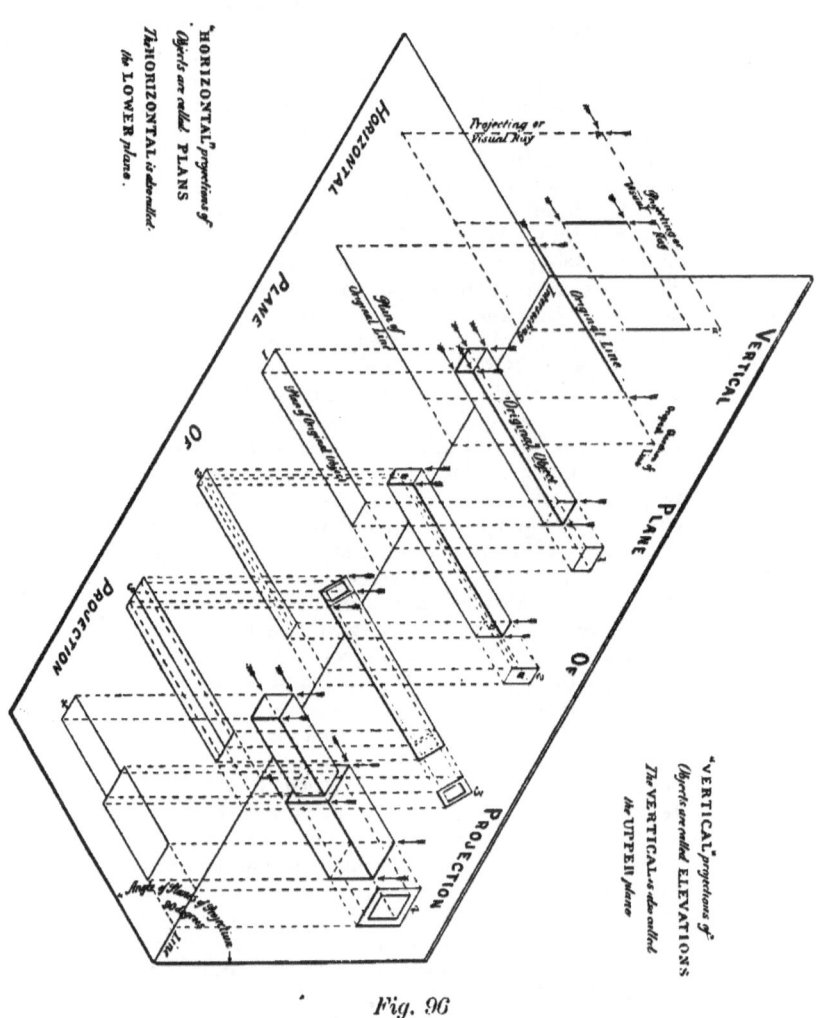

Fig. 96

end elevation is A in the VP, is obtained; for
the two *ends* of the object being vertically disp
their edges only seen from above, their horizont:
severally the lines 1', 1 ; 2', 2, for the sides, and
ends. The exact shape of the object's upper
the direction of the arrow, is truly shown in the p
of its bounding edges obtained by projection.

As the plans of the other three original o:
elevation by B, C, D, in Fig. 97, are obtained in
plan of A, all the necessary construction lines
being shown in, nothing further need here be sn
their projection, it being advisable that the stu
out for himself that it may be the better re
exercise in projection, which we will call Shee:
objects being *given*, and their plans *required*,
" Projection from the Upper to the Lower Plane "
should be drawn to as large a scale as a half-imper
paper will admit of, leaving a fair margin a
elevations A, B, C, D should be disposed in a 1
the long way of the paper, and the lengths of
assumed to be such as will occupy about two-
between the IL and the lower edge of the sheet o:

24. In our explanation of the method of c
projections of the simple objects shown in perspe
(Fig. 96), we purposely ignored the fact that *all*
were not "solid" in the sense of their having perf
prismatic in form. Our reason for this was to
the mind of the student on his first introductior
part of our subject; but as it will be necessary f
to obtain other views of objects than their me
elevation (as shown in Figs. 96 and 97), we wi:
further, explain what those views are, as they wil
for the completion of the problems given in Sheet

On referring to the object in the diagrai
projections in the VP and HP are figured 2, 2
description of it that it represents a carpenter's
composed of *two* material substances—wood 1
manufacturing such an article it would be impo
much of each of the substances which go to ma:
put into it. To decide this, we must have such a
show how far the lead or plumbago extends int:
it. Such a view is called a "section," and is ot
the pencil to be cut—horizontally, in the case befo:
the middle of its depth from end to end, the cu:
the pencil into two halves. This cutting, when tl
pencil is removed, will show at once the extent
"plane of section" will have passed through it.
of the section of the pencil after being cut wil
on. The principle involved in obtaining it is tl
plane of projection—or the sheet of drawing-pape

MECHANICAL AND ENGINEERING DRAWING

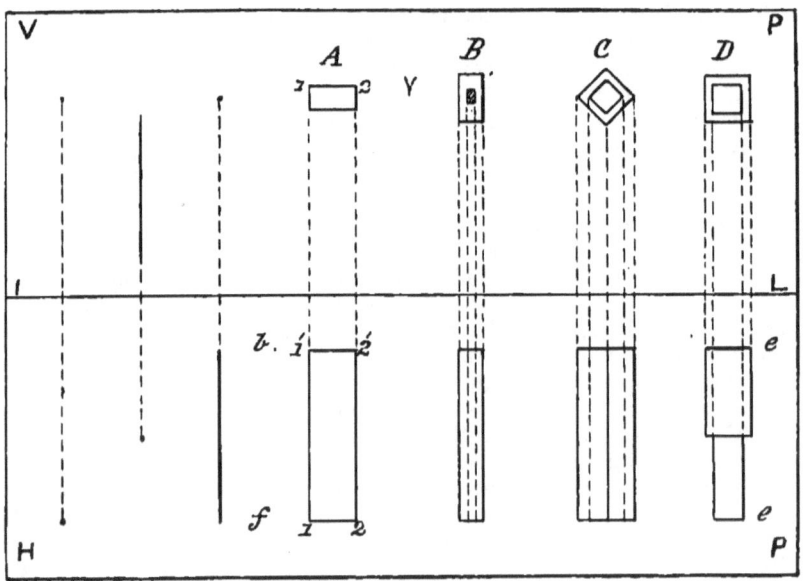

Sheet 1 Fig. 97

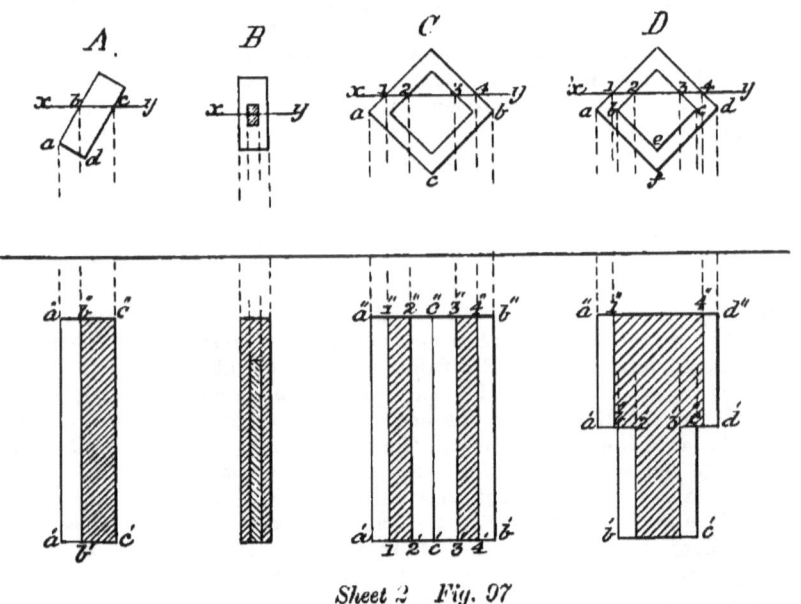

Sheet 2 Fig. 97

with the object *beneath* it, the visual rays from all its principal points being projected on that plane in points which are afterwards joined by right lines. Such a view when obtained is called a "sectional plan," the plane of its projection being the original HP, the position of the object only with respect to that plane having been reversed.

A similar plan of the object whose projections are figured 3, 3, in Fig. 96, will have quite a different appearance to that given of it in Fig. 97. It is, we are told, a *hollow* wooden tube, whose thickness is shown in its end elevation on the VP. When looked at from *above*, in the direction of the arrow, its appearance is the same as if it were merely a rectangular prism of solid material throughout, with its sides inclined to the HP: but if a cutting plane be caused to pass through it horizontally from end to end, it would then be seen (on the removal of the part cut off) that its interior is *hollow*, and its "sectional plan" something very different from that before obtained. What that view would be we shall see later on.

25. Now, in addition to the simple plan and front elevation of an object being given, it is necessary, before its exact shape and construction can be understood, to have one or more "side views" or side *elevations* of it, for there is hardly anything—in mechanical constructions, at any rate—which is of the same shape when viewed from the side and end. To show how such views are obtained, let VP and HP in the sketch (Fig. 98) represent the two planes of projection, as before, in their normal position—that is, at right angles to each other—and let it be assumed that the VP is in two parts, A and B, hinged together at *c d*, and that the part A is capable of being swung round on *c d* until it is at right angles to the part B. When so swung we have virtually *three* "planes of projection," two of which are vertical and one horizontal, and each at right angles to the other. With planes in these positions it is evident that three *different* projections of an object may be obtained on them. Let the object be, say, a simple prismatic solid, as S, and let its position be such that two of its sides are respectively parallel to the two vertical planes A and B; then the view obtained on A, looking in the direction of the arrow *s*, will be a side elevation of S; and that on B, in the direction of *f*, a front elevation; the plan P of the object in the HP being obtained as shown by the projectors to that plane. If, then, the part A of the VP, with its obtained side view of S, be swung back on *c d* into its original normal position with respect to the part B, we should have in the VP a front and side elevation of the original object S, and in the HP the plan P or view obtained when looking in the direction of the arrow *t*. On turning down the VP on the IL as a hinge until it and the HP become, as before explained, the one flat surface of the sheet of drawing-paper, the *three* true projections—viz., the front and side elevations and the plan of S —will appear as shown in Fig. 99. In that figure the assumed motion of the part A of the VP into the position of being at right angles to the part B, is shown by the dotted arcs and arrows, the winged arrow indicating the assumed motion of the plane A, and the barbed ones the transference of the projections.

MECHANICAL AND ENGINEERING DRAWING 55

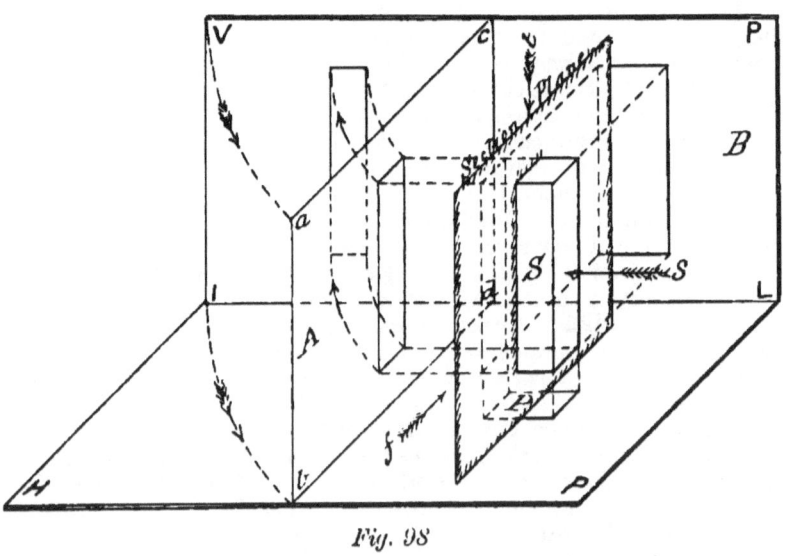

Fig. 98

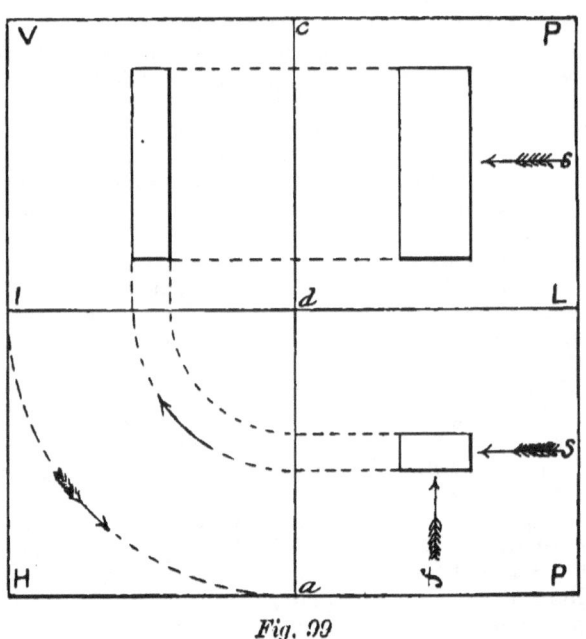

Fig. 99

56   FIRST PRINCIPLES OF

From the foregoing explanation it will be seen, on reference to Fig. 98, that a sectional front or side elevation of an object may be obtained in the same way as a simple elevation, for it is only necessary to assume the object as cut through by a vertical section plane, such as that shown in the figure, and the part cut off by it nearest the eye removed ; the view then obtained when looking in the direction of the arrow *s* would be the elevation required.

26. Having explained at some length the specific meaning of a sectional plan, and of a front, and side, and " sectional elevation " of an object, we will now revert to the problem of showing how to obtain by actual projection the sectional plans of objects, taking for our purpose those given in the diagram Fig. 96. The first on the *left* is assumed to be a beam of wood of rectangular section, shown with its two widest sides parallel to the HP, and its narrow ones perpendicular to it. As a horizontal section of a beam in such a position would give in plan a similar projection to the one already shown, and lettered in the diagram Fig. 96 " Plan of Original Object," we will, for the better practice afforded, assume the beam to have its sides inclined to the HP, as shown at A, Fig. 97, Sheet 2, and the cutting or section plane *x y* to pass through it horizontally when in that position from end to end. We will assume also that all the sides of the beam are coloured green. The problem then is to obtain by projection a plan of the beam when the upper part cut off by the section plane *x y* is removed.

Now it is evident on looking at the beam from *above* that the surface exposed by the cutting plane, showing the nature of its material, will have two edges at *b, c,* parallel to each other and at right angles to the VP, and in length equal to that of the beam ; and at *a* an edge parallel to that at *b,* the surface between these latter edges being that untouched by the cutting plane, and therefore coloured green ; the lower edge of the beam at *d,* not being seen from above, will have no counterpart in the plan, and the two ends of the beam being square with its sides, and therefore parallel with the VP, will be represented in plan by lines in that position. Therefore, having drawn in—in the VP of the sheet of paper, as at A, Sheet 2—the end elevation of the beam with its sides at the intended inclination to the HP, let fall from the points *a, b, c* projectors perpendicular to the IL into the HP, set off on the projector let fall from *a* a distance from the IL equal to that the inner end of the beam is assumed to be from the VP—which is arbitrary—and from this point *a''*, on the same projector, set off a distance *a''*, *a'*, equal to the length of the beam ; through *a''*, *a'* draw lines parallel to the IL cutting the projectors from *a, b, c,* in *a''*, *b''*, *c''*, *a'*, *b'*, *c'*; join these points by straight lines as shown, and the required projection is obtained. As the surface bounded by the parallel lines *b''*, *b'*, *c''*, *c'*, and *b''*, *c''*, *b'*, *c'*, is the plan of the section of the beam exposed after cutting, it is indicated by drawing lines with the 45° set-square, as shown.

The projection of the sectional plan of the " carpenter's pencil," shown in elevation at B, Sheet 2, will present no difficulty to the student, as the procedure is clearly shown ; there is, however, one

MECHANICAL AND ENGINEERING DRAWING   57

point in reference to it to which his attention is to be directed. The pencil being made of *two* kinds of material, this is indicated in the projection by drawing the sectional lines referred to in the last problem in opposite directions across the materials, as shown. This point will, however, be more fully enlarged upon later on.

From the projection of the pencil we pass on to that of the hollow tube, whose elevation is given in Sheet 2 in the VP at C, the section plane being shown by the line *x y*. Now in the plan of this object, given in Fig. 97, Sheet 1, it will be noted that as there are only three side edges, and the two ends, seen from above, its "plan" is obtained by the projection of these edges into the lower plane or HP, and joining them up, as shown. But in the problem before us the *upper* part of the tube is assumed to be *cut away*, leaving the section of two of its sides exposed, together with part of its interior. To find its plan under these conditions we proceed as follows :—From *a* and *b* in the elevation C, let fall projectors perpendicular to the IL into the HP ; and as the tube is assumed to be of the same length as the beam and pencil, and its *back end* the same distance from the VP as their ends are, at this distance from the IL, and parallel to it, draw the line *a″ b″*, cutting the projectors from *a* and *b* in those points ; and parallel to this line and at a distance *a″ a′* equal to the length of the tube from it draw the line *a′ b′*. We have so far obtained the bounding lines of the plan sought. For the plan of the parts of the tube cut by the section plane *x y*, let fall projectors from points 1, 2, 3, 4 in the elevation, cutting the lines *a″ b″*, *a′ b′*, in the plan in points 1′, 2′, 3′, 4′, 1″, 2″, 3″, 4″ ; join these points by lines as shown, and the surfaces between each of these pairs of lines will be the plan of the parts of the sides of the tube seen from above when the upper part cut off by the plane *x y* is removed. By its removal, however, a part of the interior of the tube is now seen when viewed from above, and this must be indicated in the plan. The part seen is the angle formed by the meeting of the two bottom inside surfaces of the tube immediately over the point lettered *c*, and as these are plane surfaces, their intersection forms a line the plan of which is found by letting fall a projector from *c* cutting the lines *a″ b″*, *a′ b′*, in *c″ c′*, and joining them by a straight line. With the section lining of the parts cut by the plane *x y*, the "sectional plan" of the "original object," or hollow tube C, is completed.

The projection of the sectional plan of the object represented in end elevation by D, in Sheet 1, would give little useful practice to the student if kept in the same position as there shown, all its surfaces being either parallel or perpendicular to the HP, and therefore resulting in a very trifling change in the plan (got by a horizontal section of it) from that previously obtained ; it is consequently shown with its principal surfaces *inclined* to the HP, thereby giving a more difficult, but more useful, problem in projection. With the original object in this altered position the student will at once be struck with the identity in appearance of its elevation with that of the hollow tube in the last problem, and he will perhaps be momentarily puzzled to understand how two apparently similar end views are the vertical projections

58   MECHANICAL AND ENGINEERING DRAWING

of two such different objects.   Here we have an instance showing
the absolute necessity of a pre-conceived knowledge of the form of
the object to be delineated, or the possession of a model of it.   Having
either, little difficulty would be experienced in fully understanding the
similarity of the two views in the diagram.   The object represented
by D in elevation (Sheets 1, 2) is merely a combination of two prisms
of different dimensions cut from one solid piece of material, from which
a portion is cut off by a plane—such as a saw-blade—throughout its
whole length, the cut being a horizontal one ; and the problem is to
show the actual appearance of the remaining part of the post when
looked at from above, *after* the part cut off is removed.

To do this, let $x\,y$ in the elevation D (Sheet 2) be the section
plane, as before ; let fall projectors from a and d into the HP, and
assuming the back end of the object to be the same distance from the
VP as those of the pencil and tube, draw parallel to the IL a line
cutting the projectors from a and d in $a''\,d''$; from a'', on the projector from a, set off $a''\,a'$, the length of the thick end of the post—
which is arbitrary—and through a' draw a faint line $a'\,d'$ parallel to
$a''\,a''$; then from b and c in the elevation D let fall projectors into the
HP, cutting the line drawn through $a'\,d'$, in $b'\,c'$, and from these last
points set off on their projectors in the points $b''\,c''$ the length of the
small part of the post ; then a line drawn through $b''\,c''$, parallel to
$a''\,d''$, will give the bounding lines of the plan of the post.   For the
sectional part of it, let fall projectors from points 1 and 4, and 2 and 3,
in the elevation D, the former to cut the line $a''\,d''$ in points 1' 4', and
the latter the faint line $a'\,d'$ in points 2' 3'; through 1' 4' draw lines
parallel and equal to $a''\,a'$; and through 2' 3' lines parallel and equal
to $b'\,b''$; the surface enclosed by the last drawn lines is that made by
the cutting plane $x\,y$, and is indicated as such by the section lining.
As the post is of solid material, the edges formed by the meeting of its
under sides in e and f will not be seen from above, and are, therefore,
not shown in the projection.

CHAPTER IX

PROJECTION IN THE UPPER PLANE

27. Having carefully worked out the problems given in Sheets 1 and 2, and studied, with the assistance of Figs. 98 and 99, and the descriptive matter in connection with them, the principles involved in obtaining the sectional plans and elevations of objects, the student should find no difficulty in solving the problems in the projection of solids which are now to follow.

The first subject we take—it being the simplest of all the plane solids—is the "cube"; but although simple, it is necessary that the student should fully comprehend the specific relations of its various faces and edges to each other, before starting to find its projections. This solid is defined as one having six equal sides or faces, all of them

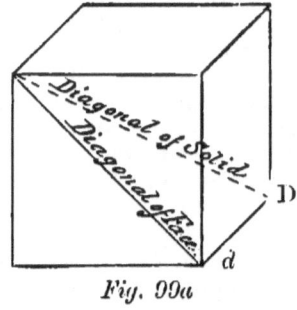

Fig. 99a

squares. It follows from this that *adjacent* sides must be at right angles to each other and *opposite* faces parallel; the adjacent and opposite edges of the solid having the same relative positions. Bearing these facts in mind, the projections of the cube are easily obtained. To put the original object on the paper as required in the problems, it is necessary to note the difference between the diagonal of a *face* of a cube, and a diagonal of the cube itself. The first is a line joining the opposite corners of any one of its faces (as *d*, Fig. 99a), and the latter is an imaginary line joining the vertices of any two of its opposite solid angles, or those formed by the junction of three adjacent

59

60   FIRST PRINCIPLES OF

faces of the solid, as the dotted line D in Fig. 99a. Our first problem
in the projection of this simple solid is— ·

Problem 30.—*Given the front elevation of a cube, with the diagonal
of one of its faces parallel to the VP, and perpendicular to the HP,
to find its side elevation, or a view of it looking in the direction of
the arrow x.* To solve it—

Draw in on the left side of the sheet of paper, with the 45° set-
square, the square A, B, C, D (Fig. 100), with its diagonal AC at
right angles to the IL, as shown. This figure will be the front
elevation of the given cube in the position stated in the problem. To
find its side elevation, through points A, B, C, draw projectors of in-
definite length parallel to the IL, and as the back face of the cube is
parallel to its front one and the VP, draw a line at right angles to the
IL, cutting the projectors from A, B, C, in a, b, c ; this line will represent the back face of the cube. Then, as the edges of a cube are all
of the same length, set off from a, on the projector drawn through it, a
distance a a', equal to the length of a side of the cube, as AB, and
through a' draw the line a' b' c' parallel to a b c ; the four bounding lines,
a c, c c', c' a', a' a', and the line b b' when joined up as shown give the
side elevation (Fig. 101) of the cube when looked at in the direction of
the arrow x. Next—

Problem 31.—*Let the cube be cut by a plane through its diagonal AC,
and let it be required to give a side elevation of it when the part to
the left of the cutting plane is removed.*

Here the part of the cube to the left of the diagonal AC being
removed, only four edges of the part left are seen. These are the top
edge from point A to the point beyond it nearest to the VP, and the
corresponding bottom edge from point C ; also the two edges of the
front and back faces of the cube, cut through by the section plane.
Therefore, in the projector drawn through A, Fig. 100, at any convenient point (say d), draw, as in Fig. 102, at right angles to the IL,
the line d e, and parallel to it, at a distance equal to the length of the
side of the cube, the line d' e' ; join the points d d', e e', as shown, and
the required projection is obtained. Again—

Problem 32.—*Let the original object—the cube—be cut by a section
plane, as SP, and a side elevation of it be required, after the part
cut off to the left of the cutting plane is removed.*

In this case the section plane cuts through two *adjacent* sides of
the cube, leaving parts of those sides, as SA and PC, in view. To
obtain the projection required, at any convenient point (say f), in the
projector drawn through A, Fig. 100, draw a line fy perpendicular to
the IL, and parallel to it at a distance ff', equal to the length of a
side of the cube, the line fy ; through the points AS, PC, and parallel
to the IL, draw the lines ff', ss', pp', gg', and the required projection
is obtained. As in Fig. 102, the section obtained is that produced by

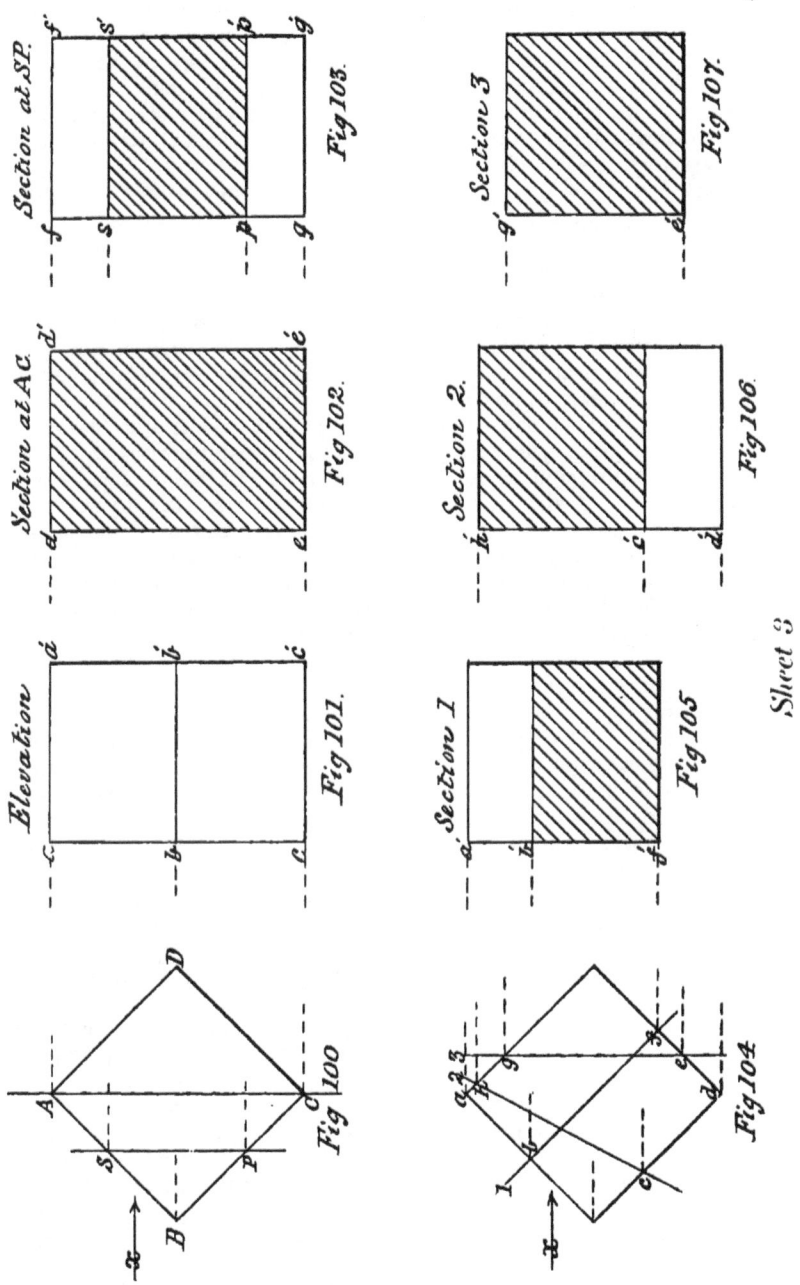

62   FIRST PRINCIPLES OF

the cutting of the cube through the plane of one of its diagonals, it is evident that no greater section could be got, and therefore its whole surface must be section-lined, as shown. In Fig. 103, however, as the cutting plane SP leaves a portion of the faces AB, BC of the cube untouched, section-lining is only required on the part of the cube actually cut, through by that plane, as shown.

28. As some further problems in connection with the projections of the cube will follow, it will be well at this point to more fully explain the significance of the lines obtained in the three preceding projections, as upon their correct comprehension depends the ease or difficulty with which subsequent ones will be found.

As in Fig. 100, the points A, B, C, D are the front ends of four edges of the cube perpendicular to the VP, and parallel to the HP, any edge produced by a section or cutting plane passing through the cube in a direction perpendicular to the VP, and making any angle with the HP, will be a line parallel to the HP, and at right angles to the lines representing in projection the two faces of the cube which are parallel to the VP. And as these faces are distant from each other a space equal to the length of a side of the cube, the projection of these faces will in all cases be lines parallel to each other at that distance apart, their projected lengths being dependent upon the angle the cutting or section plane makes with the HP.

This reasoning will be verified on applying it to the three projections given in Figs. 101, 102, and 103. In these figures the lines ac, $a'c'$; de, $d'e'$; fy, $f'y'$, are the projections of the front and back faces of the cube, the three parallel lines aa' bb', cc' in Fig. 101 being the projections of the three edges of the cube at right angles to those faces, and the equal parallel lines dd', ee' the projections of the top and bottom edges of the cube, of which A and C in Fig. 100 are the front ends. In Fig. 103 the boundary lines of the projection are the same as in Fig. 101, but the part of the cube exposed by the action of the section plane SP, is that between the parallel lines ss', pp', and has to be section-lined, as shown. Had all the faces of the cube been coloured, only those parts of the two seen when looking in the direction of the arrow x—viz., from the edge at S to that at A, and from P to that at C—would show of that colour; the surface exposed by the section being, of course, that of the material of which the cube is made. The faces of the cube parallel to the VP in Fig. 100 are, of course, in Fig. 103 seen only as lines, as at fy, $f'y'$, and as the cube is of solid material the edge of it at D, directly opposite to that at B, will not be seen in the side elevation.

As further problems in the projection of solids in the "upper plane," we give those shown in Figs. 105, 106, and 107, where the original object or cube, Fig. 104, is cut by section planes 1, 2, 3, making different angles with the HP, but all of them perpendicular to the VP. As there is really no material difference of procedure in obtaining such projections of the solid from that already so fully explained in the previous problems, it is not necessary here to go through the process in detail, as the construction lines given in connection with each figure are sufficient to enable the student to work

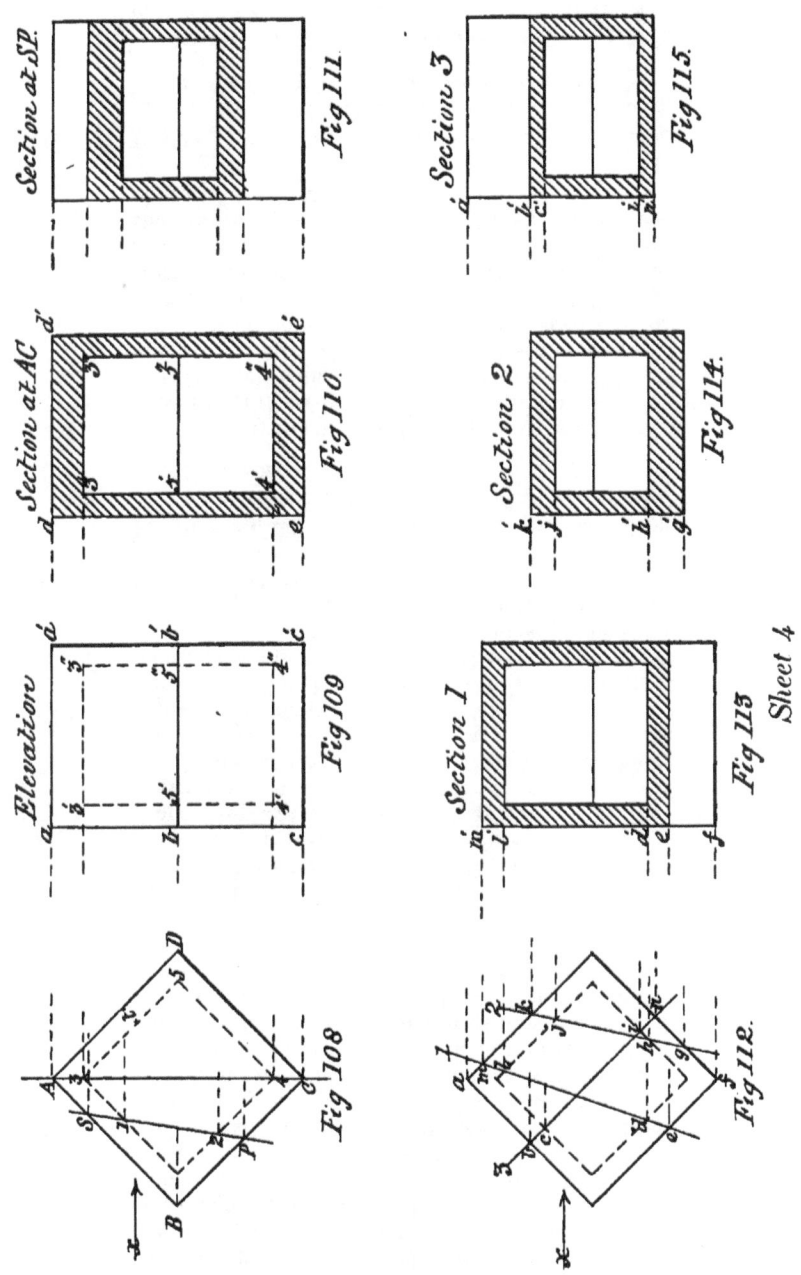

MECHANICAL AND ENGINEERING DRAWING 63

64   FIRST PRINCIPLES OF

out the problems without further explanation.
in mind, as before advised, the relationship of th
the original object—the cube—to each other, a
possible difficulty should be met with in obtainin;
tions. A couple of sheets of paper should at tl
these projections, varying the direction of the pla:
problem, so as to thoroughly master any apparent
arise in any similar problem in the future.

29. Advancing from the simple to the more d
the cube as the original object, but instead of it
out, as in the last problems, it is now hollow and
of material all over. To indicate this, in drawing
paper a front or other elevation of the object, we
use of "dotted" lines. In mechanical drawings
are invariably used to indicate those parts of an
sight, and it will be found, as we proceed in the
kind of drawing, that although appropriated m(
purpose, their use is indispensable to the draug
proper application an insight is given—on the
drawing—into the internal structure of the obj
case before us, we indicate by their use the thickn
the material of which the object is made, for, bein
appearance would be the same whether solid or l
then, as in Fig. 100, Sheet 3, the front elevation
its diagonals perpendicular to the II, and indicati
shown in Fig. 108, Sheet 4, the thickness of the n
made, we proceed to obtain by projection the vario;
Let the first be—

Problem 33.—*The front elevation of a hollow cu*
its side elevation, or a view of it when looked
the arrow x.

For convenience, letter the four corners of
cube Fig. 108 as in Fig. 100, and obtain a side
Fig. 101, lettering it in the same way. The t\
identical; but to show that the cube represented
set off from points b, b' the thickness b 5′, b 5″ of
in Fig. 108. Through 5′, 5″ draw dotted lines pa
from points 3, 4 (Fig. 108) draw projectors to
drawn through 5′, 5″ in points 3, 3′; 4, 4′; dot
the last-named points, and the required ele;
obtained. Next—

Problem 34.—*Let the cube Fig. 108 be cut by a*
its diagonal AC, and an elevation of it be r;
to the left of the cutting plane is removed.

Proceed as before to obtain, as in Fig. 102
dd', $d'e'$, $e'e'$, $e'd$ of the section; set off at 5′ [

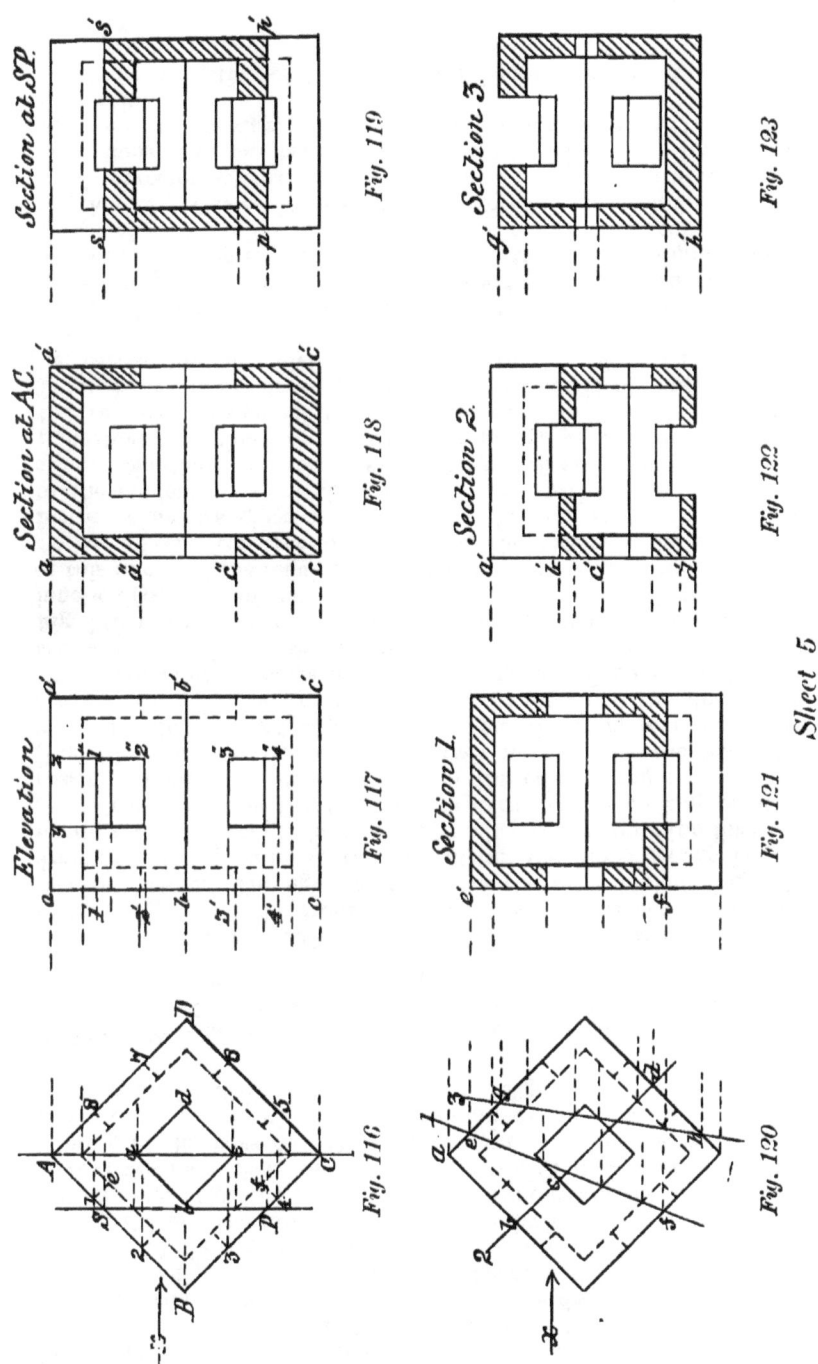

66　　　　　　　FIRST PRINCIPLES OF

front and back faces of the cube—the thickness of its sides, and through 5′ 5″ draw lines parallel to those sides; through points 3, 4 in Fig. 108 draw projectors to cut the parallels drawn through 5′ 5″ in points 3′ 3″, 4′ 4″; the surface included between the inner and bounding lines of the figure is that exposed by the cutting plane AC in passing through the cube, and is indicated by cross or section-lining. As the removal of the part of the cube to the left of the cutting plane AC exposes its *interior*, the angle formed within it at point 5, by the meeting of the sides AD, DC, will be seen as a line between points 5′ 5″ in Fig. 110. With the drawing in of this line as shown, the required sectional elevation is complete. Again—

Problem 35.—*Let the cube Fig. 108 be cut by a plane SP, and let it be required to give a side elevation of it, when the part cut off to the left is removed.*

Proceed as in Fig. 103, Sheet 3, to obtain by projection the top and bottom edges, and the front and back faces of the cube, and letter them as before. Through points S P in the section plane Fig. 108, draw projectors parallel to the IL, cutting the line fg (Fig. 111) in points sp, and through these points parallel to ff' draw the lines $s's''$, $p'p''$. Then to show the thickness exposed by the cutting plane in passing through the sides AB, BC of the cube, through points 1, 2 (Fig. 108), and parallel to the IL, draw in Fig. 111 the lines 1′1″, 2′2″. Parallel to, and at a distance from the front and back faces fg, $f'g'$, equal to the thickness t of the cube, draw the lines 1′, 2′; 1″, 2″: cross-line the surface between the inner and bounding lines of the section as shown, and as in the case of the section obtained in Fig. 110 the interior of the cube is exposed, a line will be seen at the junction (point 5) of its two inner inclined surfaces; with the drawing of this line 5′ 5″ the required sectional elevation is complete.

As all the construction lines and the reference letters in the further examples given in Figs. 112 to 115 are shown in, it is left to the student to work them carefully out without further explanation. No difficulty need be experienced with either of the sectional projections, provided due thought is given, as before advised, to the relative positions of the faces and edges in the original object. Throughout the problems, the views required are in all cases those of the part of the object left, when that to the left of the cutting plane is removed.

30. Having satisfactorily worked out the problems in Sheet 4, the student will now be able to proceed with the following, which will require on his part a closer study of the construction of the original object than before, for although it is still in the form of a cube, the perforation of its sides by openings, as shown in Fig. 116, Sheet 5, will involve greater attention to the method of procedure in obtaining its projections than has before been necessary. This, however, is to be expected in drawing, as in every other art worthy of study or acquisition.

On an inspection of Fig. 116, it will be seen that the cube is in the

MECHANICAL AND ENGINEERING DRAWING   67

same position with respect to the VP and HP as in the two last sheets of problems—that is to say, the diagonals of its front and back faces are respectively perpendicular and parallel to the HP, the faces themselves being parallel to the VP. Each side of the cube, however, instead of being, as in the last problems, solid, has now a square hole through it, the sides of the holes being parallel to the sides of the cube. It is therefore possible to see right through the cube from any of its six sides. The consideration of these simple facts in connection with the original object will be found of service in the attempt to obtain any required projection of it. Let the first be—

Problem 36.—*Given the front elevation of a hollow cube, with square openings in a central position in each of its sides, and a diagonal of its front face perpendicular to the HP, to find its side elevation when viewed in the direction of the arrow on the left.*

First draw in the front elevation of the cube, showing the thickness of the material by dotted lines, as in Fig. 108, Sheet 4. Letter the corners of it A B C D as before. Divide each of the edges AB, BC, of the cube into three equal parts, in the points 1, 2, 3, 4, and through these points draw faint lines parallel to the edges BC, AB respectively across it. The square *a b c d* formed by the intersection of these four lines gives the opening on that face of the cube, and the one directly behind it, and the position of those on the other four faces is shown by the parts of the same four lines which cross the thickness of the sides of the cube at the points 1 to 8 in their edges. On putting in the square *a b c d* in full lines, and dotting in those last referred to, the elevation of the cube shown in Fig. 116, Sheet 5, is that specified in the problem.

To give the side elevation required, proceed to obtain first, as in Fig. 109, an elevation of the cube *without* openings in its sides. Then, to show the openings that will be seen when looking in the direction of the arrow *x*, we have to remember that only two faces of the cube (of which AB, BC, are the front edges) are in sight, and that therefore only the two openings figured 1, 2 ; 3, 4 are seen. Now, as the sides of these openings are one-third the length of the sides of the cube, and as the four edges of these openings figured 1, 2, 3, 4 will be seen of their *actual* lengths, to show them in the position they occupy on the cube, set off in points *y z* from the lines *ac*, *a'c'* (Fig. 117)—the front and back faces of the cube—the distance (*a* 1 or *b* 3) that the edges of the square opening *a b c d* are from the edges of the cube. Through these points draw short lines in the upper and lower half of the figure parallel to *ac a'c''*, and from points 1, 2, 3, 4 (Fig. 116), and parallel to *aa'* or *cc'* (Fig. 117) draw lines cutting those drawn through *y z* in 1' 1'', 2' 2'', 3' 3'' 4' 4''. The *outer* edges of the two openings seen from *x* will thus have been obtained. Two of their *inner* edges, one at *e* and another at *f*, are also seen, the first at *e* being that of the top side of the upper opening, and that at *f* the corresponding edge of the lower opening. Projecting these points over on Fig. 117, and drawing in the lines they represent, will give their elevation. To complete the view of

68   FIRST PRINCIPLES OF

the cube required, we have yet to show the position of the two openings in its front and back faces, represented in Fig. 116 by the square *a b c d.* This is found by drawing in in dotted lines in Fig. 117, the projection of the two corners *a* and *c* of that square; the other corners *b* and *d* being directly behind the edge of the cube at B, and coinciding with it, will not be seen in the required elevation.

Figs. 118 and 119 are sectional side elevations of the same original object, the first being that of the cube cut by a plane passing through ts diagonal AC, and the second by a similar plane at SP parallel to that diagonal. With the foregoing explanation, showing how the outside view of the cube is obtained, and the assistance of Fig. 110, together with that given by having all the principal projectors shown in for each figure, the attentive student should be able to obtain without further aid or assistance the two elevations of the cube shown in the figures referred to.

In Fig. 120 is given an elevation of the same hollow cube in the same position as that shown in Fig. 116; but the sectional side elevations of it required by planes cutting through it at 1, 2, 3, and shown in Figs. 121, 122, and 123, although found in exactly the same way as Figs. 118 and 119, necessitate much closer attention on the part of the student in obtaining them, the section planes being purposely drawn in such directions as to make the resulting projections a test of his ability in applying the principles which have previously been so fully explained to him.

In obtaining the three sections required, the only likely difficulty to be met with may possibly occur at that part of each projection where the cutting plane crosses an opening in one or other of the sides of the cube. No. 1 section plane, it will be noticed, is shown to cut three such openings—viz., the one immediately in front and its fellow one at the back of the cube, and the lower opening at *f;* No. 2 plane cuts through the upper left-hand opening at *b,* the lower right-hand one at *d,* and the front and back ones, dividing the cube into two equal parts; and No. 3 plane cuts through the upper right-hand opening at *g,* and across the corners of the front and rear openings.

31. In showing the parts of the sides of the cube in section in the three views, care must be taken to note especially where the section plane enters and leaves the solid parts of the object, and crosses the open parts, for these points determine what parts are in section and seen, and what are not. In Fig. 121 the *upper* part of the projection is shown as being solid right across, because the section plane cuts through a solid part of the cube's side. The same is seen in the *lower* part of Fig. 123; but in the opposite parts of each of these figures the section plane has cut through an opening in that side of the cube, and this is indicated by the gaps shown. The same reasoning will explain the projection obtained at Fig. 122, only in this instance the section plane has equally divided the four openings it passes through. With the foregoing explanation, the three sectional elevations of the original object should be obtained without difficulty, projectors being shown in the front elevation of it (Fig. 120), partly drawn in, from all the important points cut through by the section planes.

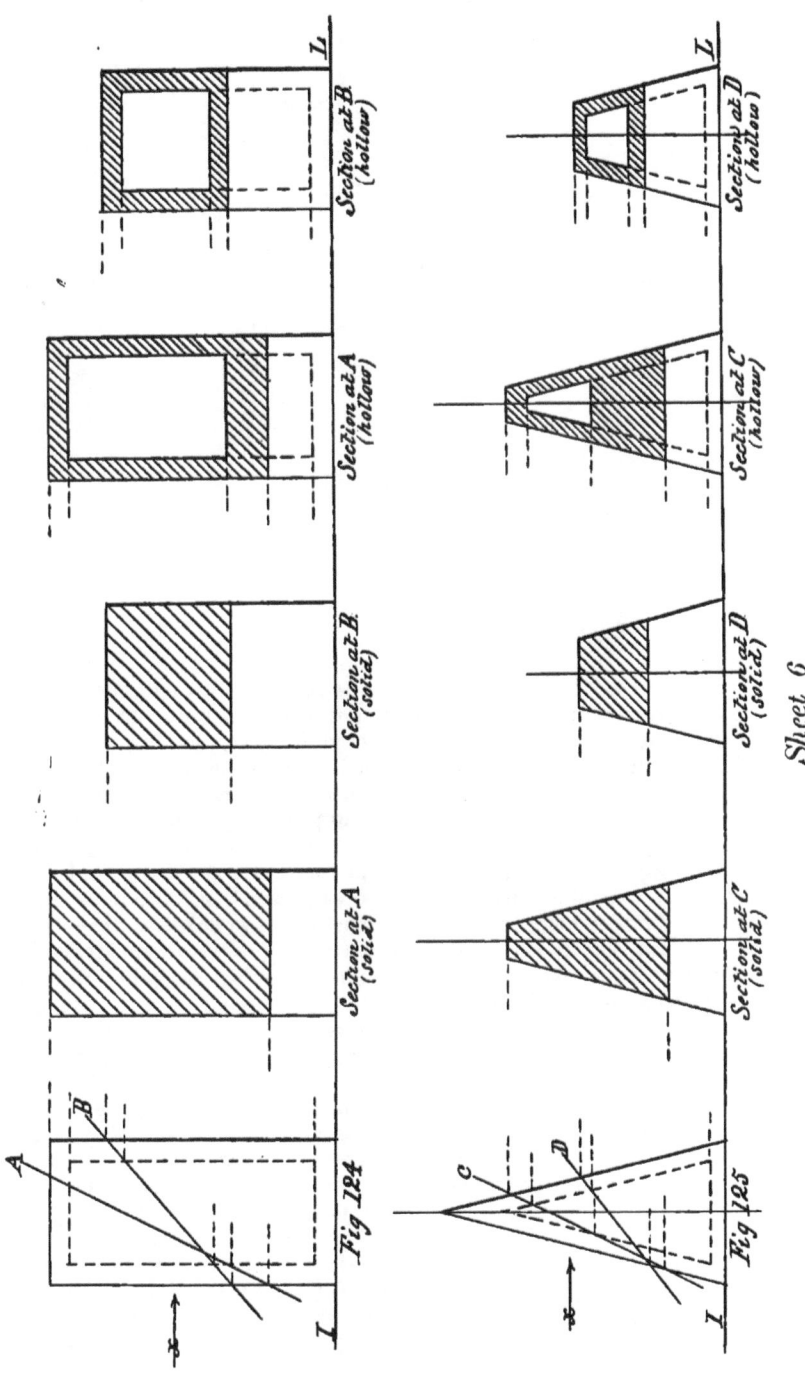

Sheet 6

70   MECHANICAL AND ENGINEERING DRAWING

As the side elevations of the prism and the pyramid are obtained in the same way as those of the cube—so fully shown in Sheets 3, 4, and 5,—a few problems in their projection are given in Sheet 6, for the student to solve without further aid than that afforded by the construction lines shown in the diagrams.

Fig. 124 is the elevation of a square prism, assumed to be solid in the first instance, with its base resting on the HP, and two of its sides —adjacent ones—respectively parallel and perpendicular to the VP. It is required to give side elevations of the prism—looking in the direction of the arrow x—when cut by the section planes A and B. Then, assuming the prism to be hollow, but with closed ends, as shown by the dotted lines in Fig. 124, side elevations of it are required when cut by the same section planes as before.

Fig. 125 is the elevation of a square pyramid, with its axis vertical, and two adjacent edges of its base respectively parallel and perpendicular to the VP. Side elevations—as in the previous problems—are required of the sections produced by the cutting planes C, D : first, assuming the pyramid to be solid, and then hollow. In obtaining the projections of the pyramid, the student must not forget that as each of its sides are triangles, their width at *any* height from the base varies with that height, and must be found accordingly.

After some further practice in "projection in the upper plane," by devoting a sheet or two of paper to finding the projections of the cube, prism, and pyramid, cut by planes drawn in other directions than those given in the problems, the student will be enabled to enter upon the next stage in advance in our subject.

CHAPTER X

PROJECTION FROM THE LOWER TO THE UPPER PLANE

32. It has already been shown in Figs. 70, 71, that if a point in the HP be the *plan* or horizontal projection of a *line*, then the line is a straight one, perpendicular to the HP and parallel to the VP, and that its "elevation" is obtained by drawing through the given point a straight line in the VP, perpendicular to the IL, of a length equal to that of the line represented by the point. It is also shown in the same figures that if a line be represented in plan—or in the HP—by a line of the same length as its original, perpendicular to the IL or VP, then the elevation of that line will be a point in the VP, where the foot of the projector drawn through the original line touches the VP. These two cases embrace the principles involved in finding by projection the "elevation of an object when its plan is given," or projection from the lower to the upper plane.

Having previously worked out the problems of finding the "elevations" of any of the four plane geometrical figures from their "plans," we proceed now to show how the elevations of solid objects, whose sides or faces are plane figures, are obtained when their plans are given. As previously advised in the case of the cube, prism, etc., no difficulty will be met with in obtaining these projections, if the relations of the several sides of the original objects to each other are previously understood. Our first problem in this part of the subject is—

Problem 37.—*Given the plan of a rectangular slab of solid material, with its vertical sides inclined to the VP, to find its elevation.*

Let the rectangle ABCD, Fig. 126, Sheet 7, be the plan of the slab in the position stated in the problem. Thus shown, all its sides and ends are rectangular plane surfaces, the upper and under ones being assumed to be parallel to the HP. The parts of it that will be seen, looking in the direction of the arrow x, will be the end AB and the side BC. The points A, B and C are the *upper* ends of the corner edges, or lines, which have a length equal to the thickness of the slab; therefore to find its elevation, through points A B C, draw projectors Aa, Bb, Cc perpendicular to the IL into the upper plane or VP. If the

71

72   FIRST PRINCIPLES OF

slab is resting *on* the HP, set off its thickness on the projector from A, from the IL, as at *a'*, and through *a'* draw a line parallel to IL between *a'* and *c'*; then Fig. X will be the elevation required.

But if the slab is not resting on the HP, but is assumed to be some distance above it, then on the projector from A set off from the IL the assumed distance that the *top* surface of the slab is from the HP, say at *a*; through *a* draw a line parallel to the IL cutting the projectors from B C in *bc*; then *a b c* will be the elevation of the points A B C in the plan, and the line drawn through them that of the *top* surface of the slab. Set off from *a*, on its projector towards the IL, the thickness of the slab; through this point draw a line parallel to *a b c*, or the IL, and Fig. Y will be the required elevation. The corner D of the slab will of course not be seen in the elevation; but to show that the form and position of the slab are understood by the student, he should indicate its position by a dotted line, as shown.

Our next object is that of a solid with inclined sides, and the problem is—

Problem 38.—*Given the plan of a frustum of a square pyramid with its base on the HP, and its base edges inclined to the VP; to find its elevation.*

Let Fig. 127 be the plan of the frustum in the position specified, A B C D being the four corners of the base, and let its height be equal to the length of the side *ab* of its upper surface, shown in plan. Find the elevation of the points A B C in the plan, by projectors to the IL, cutting it in points A', B', C'. On the projector from A produced, set up from the IL the height of the frustum in point *a''*, and through it draw indefinitely a line parallel to the IL; then the points *a', b', c'* in this line, where the projectors from *a, b, c* in the plan cut it, will be the elevations of the three corners of the upper face of the frustum, seen when looking at it in the direction of the arrow *r*. Join *a'*A', *b'* B, *c'*C' by straight lines, as shown, and the required elevation is obtained.

Proceeding from the simple to the more difficult, our next problem is—

Problem 39.—*Given the plan of the frustum of a square pyramid resting on the HP, and surmounted by a cube, its sides being inclined to the VP; required its elevation.*

Let Fig. 128 be the plan of the combined solid, in the position given in the problem, and let the height of the frustum be equal to the length of a side of the cube. Then, having found the elevation of the frustum, as in the last problem, find by projectors from points 1, 2, 3 in the plan the elevation of the three corners of the cube, seen when looking in the direction of the arrow. On the projector from point 1 in the plan set off from the upper face of the frustum, or the line *a' b' c'*, the height of the cube, and through the point 1' draw the line 1' 2' 3', and the required elevation is obtained. The elevation of any of the regular plane solids from their plan is found in the same way. For example—

MECHANICAL AND ENGINEERING DRAWING 73

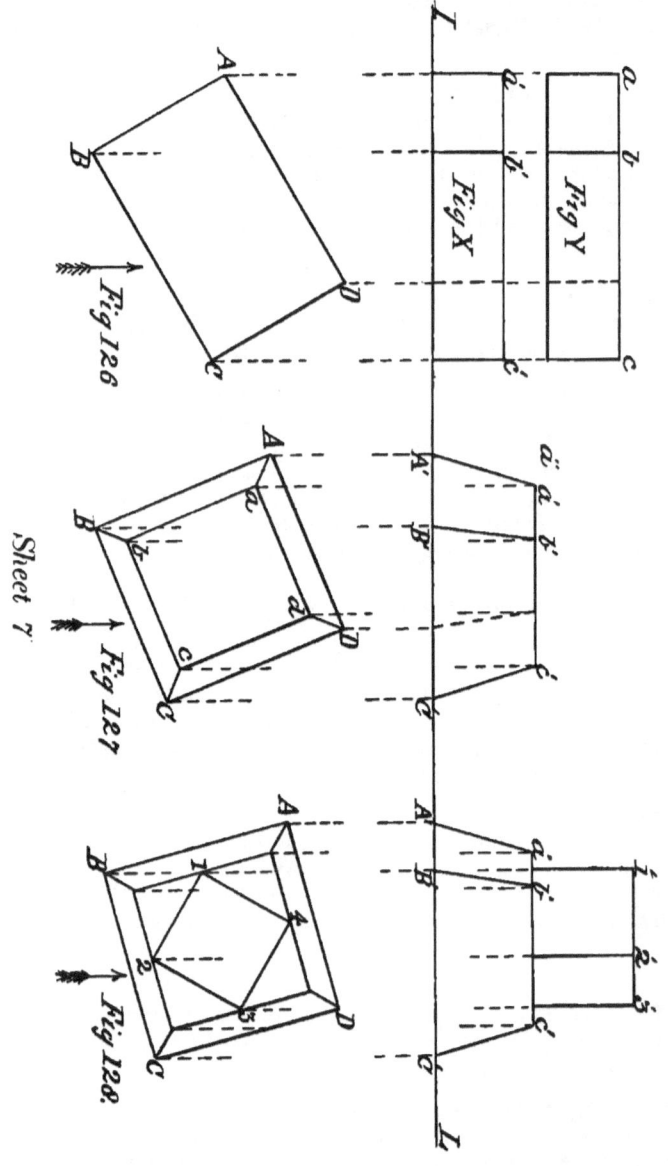

74   FIRST PRINCIPLES OF

Problem 40.—*Let Fig. 129, Sheet 8, be the plan of an hexagonal prism
with its axis vertical and its base on the HP, and let its height be
equal to twice the diameter of the inscribed circle of its base; required
its elevation.*

In the position in which the object is standing with respect to the
VP, three of its sides and four of its vertical edges will be seen. Find
the elevation of these edges by projectors through the points A B C D;
on the one through A set up the height of the prism equal to twice a',
and at this height and parallel to the IL, draw a line cutting the
projector from D in d, and the required elevation is obtained. Again—

Problem 41.—*Let Fig. 130 be the plan of a square pyramid, its axis
vertical, its base resting on the HP, and its height equal to twice
the length of the diagonal of its base; find its elevation.*

As the solid is resting with its base on the HP, the elevation of its
three base corners that will be seen—viz., A B C—will be found in
$a\ b\ c$ on the IL. Its apex is the point p in the plan where the two
diagonal lines—which are the plans of its side edges—intersect. Find
the elevation of the axis by a projector through p; set off on this from
the IL upwards the height of the pyramid in the point p', and join p'
by right lines with points $a'\ b'\ c'$ on the IL, and the required elevation
is obtained.

The "sectional" elevation of an object is obtained from its plan by
similar methods.

Problem 42.—*Let Fig. 131 be the plan of a solid cube resting
with one of its faces on the HP, and let it be cut by a plane SP
perpendicular to the HP, and an elevation of it be required, when
the part X, cut off by the plane, is removed.*

Projectors being drawn from the points A S P C, into the upper
plane, as shown, and the height of the cube set off from the IL in point
a'; on the one drawn through A, a line through a', parallel to the IL,
cutting the projector from C in c' and the cross-lining of the part cut
through by the section plane, completes the elevation required.
Again—

Problem 43.—*Let Fig. 132 be the plan of a hollow square pyramid,
its height being twice the length of a side of its base, and a sectional
elevation of it on the line SP be required.*

As the cutting plane passes through the axis of the pyramid, its
section will be a triangle. First find the elevation of this triangle,
making its altitude the given height stated; then to show the thickness
of the material cut through by the plane, find the elevation of the
points 1, 2 in the plan on the IL, and through these points 1′ 2′, and
parallel to the sides of the triangle, draw lines meeting in the axis in

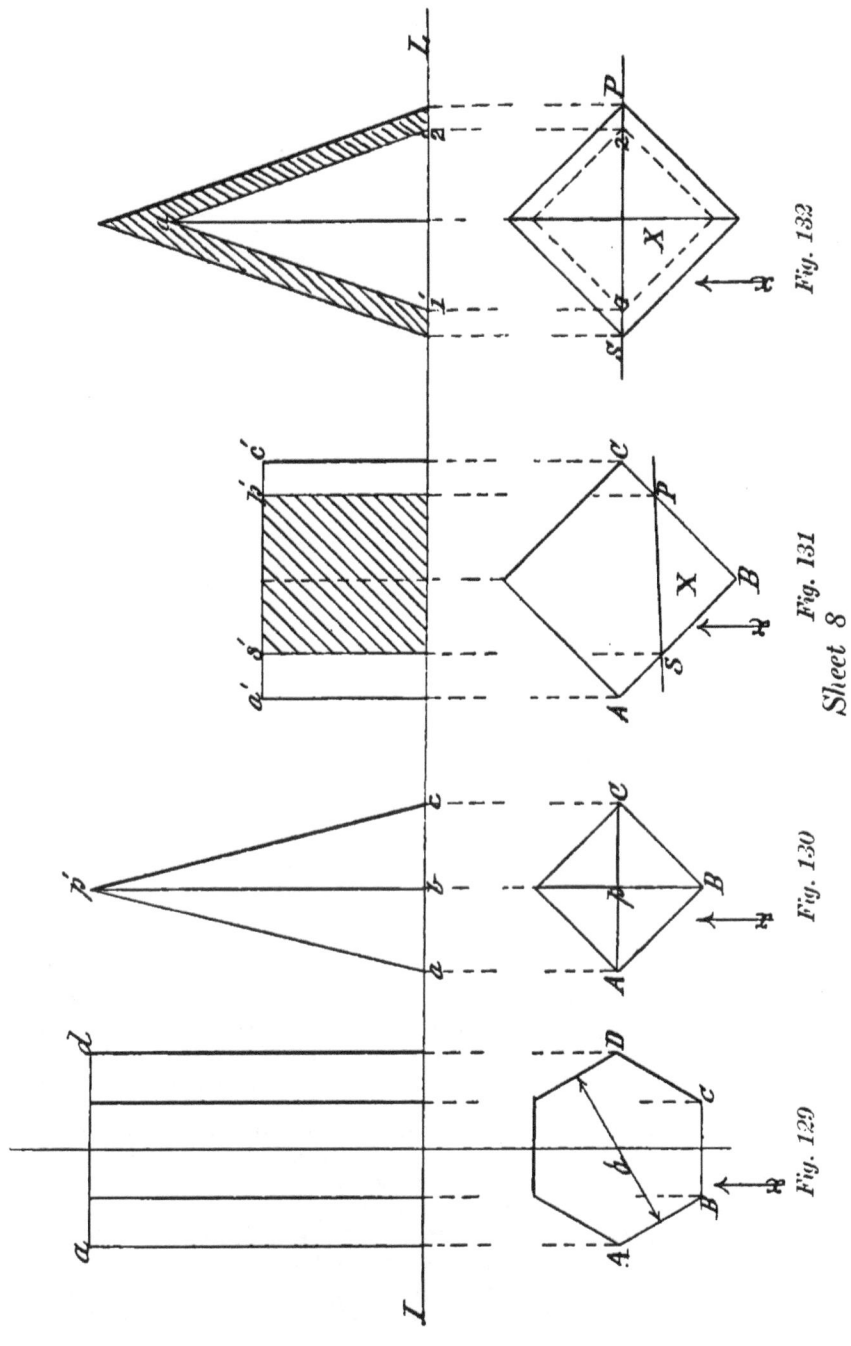

Fig. 129 Fig. 130 Fig. 131 Fig. 132

Sheet 8

MECHANICAL AND ENGINEERING DRAWING

75

76   MECHANICAL AND ENGINEERING DRAWING

the point a, cross-line the part cut by the section plane as shown, and the required sectional elevation is obtained.

From the foregoing problems it will be seen that to obtain the elevation of an object from its plan alone, without other important data, would be impossible, as the heights and conformation of its different surfaces not seen in the plan must be known before its correct elevation can be attempted.

It being necessary, at this stage of the student's progress in the study of our subject, that he should know something about the correct lining-in of his work in ink, we shall next explain the proper application of the different kinds of lines used in mechanical drawings, that he may be able to practise in spare moments on the sheets of drawings in pencil the results of his study, which it is assumed he has preserved.

CHAPTER XI

LINING-IN DRAWINGS IN INK

33. ALTHOUGH the student, up to the present stage in his study, has not been called upon to draw anything to scale—which necessitates a greater amount of exactness in the use of his pencil and instruments than he may yet have exercised—he should still have acquired a sufficient ability in their manipulation to enable him to put in fairly sound and fine lines when necessary. But as the permanence and practical use of a drawing—especially one of any engineering subject—are matters of necessity and great importance, it must be committed to paper in a better medium for its preservation than that offered by the use of plumbago. The lead-pencil is only employed for the rapid committal to the paper of the ideas embodied in the drawing, but for the preservation of these, and for constructive purposes, the design must be fixed in some coloured pigment or ink. As previously stated at the commencement of this work, China or India ink is the special pigment used by the mechanical draughtsman for this purpose, and it is now intended to explain the proper application of the different kinds of lines used in inking or lining-in an outline mechanical drawing.

We may state at the outset in this part of our subject that in what are known as ordinary, workshop, or "shop drawings," only *one* kind of line is used in inking them in, and that is a firm, sound, black line, about $\frac{1}{64}$ of an inch in thickness. In all good modern workshops, no workman is allowed to decide by measurement with his rule the dimensions of any part of a piece of mechanism shown on a drawing, as all such parts are, or should be, dimensioned with figures on the drawing itself before it leaves the drawing office. But for the purposes served by a drawing in outline, or one not intended for shop use, different thicknesses of lines are necessary to enable it to be properly read and understood. These lines are generally of three degrees of thickness, and are defined as *fine, medium,* and *shadow* or thick lines. Their use, however, without some well-defined rule of application, would be futile; as the very reverse effect of that intended would be produced by the incorrect use of either of them.

34. Now, as the proper application of these lines is directly concerned with the effect caused by light falling on an object, it is a matter of importance in this special kind of drawing that a uniform rule be adopted with respect to the *direction* in which the light is supposed to fall upon the object represented. With the free-hand draughtsman or artist, this direction is optional, as he can adapt it to the way he thinks most conducive to effect in showing up any particular object in his picture ; but with the mechanical draughtsman, as his drawings are not representations of objects as they appear to the eye, but are projections obtained by parallel rays from all parts of them falling upon certain planes having definite relations to each other, but represented by his sheet of paper, he has to adopt some rule of illuminating the visible surfaces of his objects, in accordance with the system he uses in projecting their outlines.

Although the illuminant is the sun, and its light is diffused equally around, it is generally assumed that we *see* objects by light coming from *above* and *behind* us ; but it is evident that if the light shone directly from behind, the spectator would be in his own light, and part of the object would be in shade. The light must then be assumed to come either from the right side or the left. As a rule, the rays of light are always assumed to come over the *left* shoulder of the draughtsman in *parallel* lines, and to strike the planes of projection—or the VP and HP —of the drawing at an *apparent* angle of 45° with the IL, or intersecting line of those planes. The *actual* direction of the rays is graphically shown in the diagram Fig. 133.

Let VP and HP represent the two planes of projection, and the line IL the intersecting line dividing them. In these planes draw in ABCD to represent the elevation, and a'b'c'd' the plan of a cube. In the position shown, the front and back faces of the cube are parallel to the VP, and all the others perpendicular to it. Through C in the elevation, and c' in plan, draw lines EC, ec', making angles of 45° with the IL ; then EC and ec' will represent the plan and elevation of a ray of light and the *apparent* direction in which it falls upon the VP and HP. The *actual* direction, or path of the ray, is from the upper anterior or front corner of the cube at A, to the lower posterior or back corner of the cube behind C. In other words, the ray of light is assumed to travel in a direction coinciding with the diagonal of the cube, drawn between point A and the point beyond C.

To find the actual angle that this ray of light makes with the planes of projection : At point A in the elevation of the cube erect at A, a perpendicular to AC ; on this set off from A a length Ad equal to a side of the cube ; join d and C ; then the angle ACd is that made by the ray of light with the planes of projection VP and HP. For if the right-angled triangle dAC be supposed to turn on its base AC as a hinge until its plane coincides with AC, then the angular point d will coincide with A, and the hypothenuse dC of the triangle dAC will become, as before stated, the path of the ray of light. The angle made by this ray with the VP and HP will be found to be, both by measurement and calculation, one of 35° 16″.

To make the rule—adopted by the mechanical draughtsman—as to the assumed direction of the light falling on a body still more clear, let the student cut out in stiff paper a model, allowing a strip for gumming, as in sketch A, of the right-angled triangle dAC, Fig. 133 ;

MECHANICAL AND ENGINEERING DRAWING

Elevations

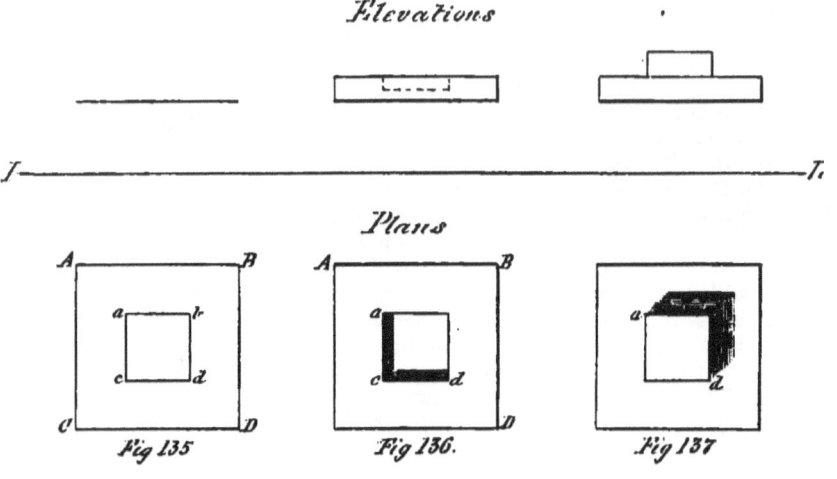

Plans

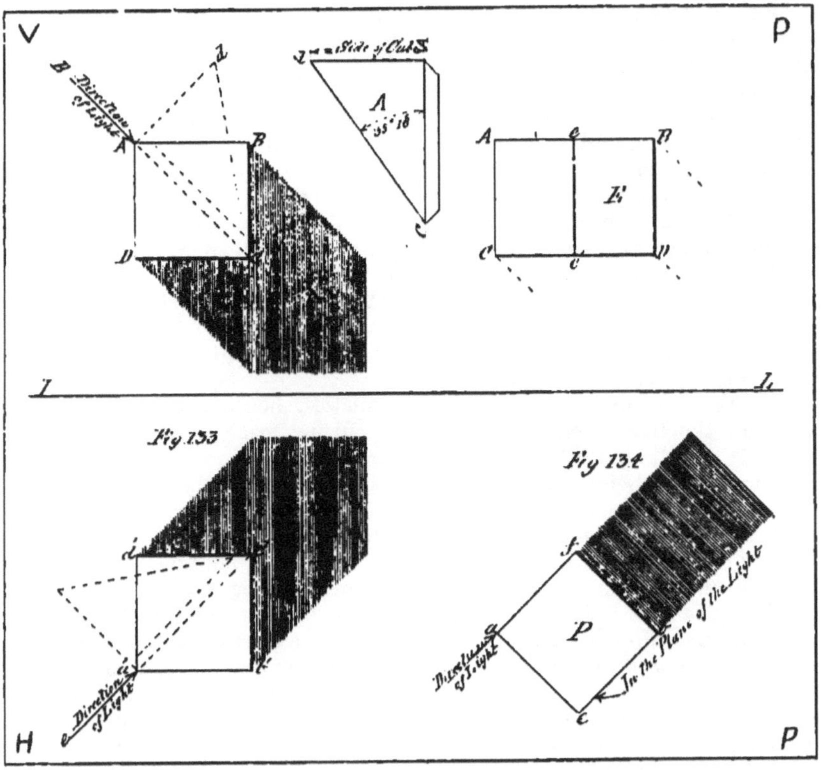

Fig 135   Fig 136.   Fig 137

80 FIRST PRINCIPLES OF

fold over the strip on the model, on line AC, until it is at right angles to the triangular part, gum the strip on its under side and apply the model to the *elevation* of the cube, making the lines AC on each coincide. Then prepare a similar model of the triangle and apply it to the *plan* of the cube on the line *a′ c′*. With the sheet of paper on which the plan and elevation of the cube are drawn, *flat* on the drawing-board, the triangular models will both stand perpendicular to the plane of the paper. If now the VP be bent over *forwards* on the IL as a hinge until it is at right angles to the HP, the direction of the ray of light with respect to both planes of projection will at once be seen, for the hypotenuses of the triangles with which the ray of light coincides are parallel to each other, and make equal angles with the VP and HP. It will also be proven by the aid of the model, that a line drawn in the VP and HP at an angle of 45° with the IL, correctly represents the elevation and plan of a ray of light when making an angle of 35° 16″ with those planes.

35. We have now to explain why a distinction should be made in the *quality* of the lines used in inking-in an outline mechanical drawing. It will be readily understood on looking at any illumined object that there are some boundaries in its surfaces on which the light will fall *direct*, and others on which it will not so fall, but will cause them to cast a shadow on some other surface beyond them. Of the former, some are often so much in the light that the edges bounding the surfaces are almost imperceptible; all such must of necessity be represented by *fine* lines. Those on which the light does not fall direct, but which cast a shadow on some more distant surface, are represented by *thick* lines. A *medium* thickness of line is only used under exceptional conditions of position of the bounding edges of surfaces; but of their use we shall speak later on. For the present we shall deal only with *fine* lines, and *thick* or shadow lines, and their application to plane-surfaced solids, or such as the student has already drawn-in in pencil. Knowing all the *forms* of the solids which have been taken as subjects for projection, their proper inking-in should present no difficulty. All that has to be determined in connection with the pencilled-in lines is what each represents—that is to say, is it an edge or bounding line on which the light falls *direct*, or does it cast a shadow? This decided, the line can be drawn-in in ink.

Taking the two projections of a cube—given in the diagram Fig. 133 as an explanatory example—let it be determined how they should be inked in. To begin with the elevation, and knowing the direction in which the light falls upon the object, it is evident that on two of its front edges—AB and AD—the light is falling direct; their representative lines—AB and AD—must therefore be *fine* ones. It is equally obvious that the edges BC and DC will cast a shadow on the VP against which the cube is resting, therefore the lines BC and DC, representing these edges, must be *thick*, or shadow lines. Three other edges of the cube are also affected by the light—viz., those of which A, B and D are the front ends; but this cannot be indicated in the elevation. The cube whose plan is shown in the diagram is for explanatory purposes set forward some distance from the IL, and must not

MECHANICAL AND ENGINEERING DRAWING   81

therefore be taken as that of the cube given in elevation.   In this
case it is assumed to be standing with one of its faces on the HP.   In
that position two of its top edges—$a'd'$ and $a'b$—are directly in the
light, and will therefore have to be *fine* lines, while the corresponding
opposite edges $b'c'$ and $c'd'$, as they will cast a shadow on the HP,
must be *thick*, or shadow lines.   Of the "cast shadows" thrown on
the VP and HP by the cube in the positions shown, we shall speak
later on.

36. To show the importance of *correctness* in the use of fine and
thick lines in an outline drawing, we give in Fig. 135 a very simple
drawing which may be made, by a different application of such lines, to
represent totally different objects.   The figure being drawn-in altogether
in fine lines may represent anything; say a square *thin* plate of any
material, with a small square lined-in on its surface.   To give an
idea of substance or thickness in the plate, this would be effected
by thickening the lines from A to B and B to D, as in Fig. 136.   The
inner square would represent a "recess" in the plate by thickening the
lines ac and cd, as in Fig. 136; but if the *opposite* sides, ab, bd, of this
small square were thick lines, then this square would represent a
"projection" on the surface of the plate or slab as shown in Fig. 137.
That Figs. 136 and 137 are plans or horizontal projections of objects is
evidenced by the position of their shadow, or thick lines, the shadows
cast by their projecting edges—shown in—indicating the direction in
which the light falls upon them.   The original drawing, Fig. 135, may
be either a plan·or elevation, there being nothing in its lining to show
which is intended.   The student should be careful to note that even in
this simple drawing of only eight lines, the *reversing* of the positions
of the thick lines bounding two of the sides of the inner square
produces not only a different appearance in the object, but gives quite a
different reading as to its construction.

37. From the foregoing explanation of the use and application
of a fine and thick line in defining the meaning of an outline drawing,
it will not surprise the student to find that an alteration in the *position*
of an object with respect to the light falling upon it may, and possibly
will, affect its lining-in.   To show the effect of such an alteration
in position, let E and P in the diagram Fig. 134 be the elevation
and plan of a cube, with four of its faces vertical, and making
angles of 45° with the VP.   In this position the light falls wholly and
directly upon the vertical face Ae Cc', leaving its parallel and opposite
one wholly in shade, while the face eB e'D and its opposite one are
in the plane of the light.   In this case the lining-in of both plan and
elevation of the cube becomes quite different to that of the same objects
when differently placed with respect to the light falling upon them.

As in the previous figure, in elevation, the under side of the cube is
assumed to be some distance *above* the HP, but in plan it is supposed
to have that side resting *upon* it.   The correct lining-in of the two
views in accordance with their altered position will then be as
follows :—In elevation, the top front edges Ae, eB, forming one
line, being directly in the light, must be a fine line.   The vertical
edge AC, being "in the plane of the light," must also be a fine line.

G

The edge *eé* is in the plane of the light, but it being so near to becoming one that would either cast a shadow or be fully in the light, a *medium* thickness of line would be most appropriate. As the vertical edge BD of the cube would cast a definite shadow on any surface behind it, it must of course be a thick line. The same applies to the two lower front edges of the cube C*e' e'*D, forming one line, which must also be a thick one. Practice among mechanical draughtsmen varying as to the appropriate use of a medium thickness of line in a drawing, it may be remarked in the case here presented that if "colour" were used to assist the eye in determining the form of the object represented in elevation by E in the diagram, the tint would be *dark* near the edge *eé'*, and gradually lighten off as it approached the edge BD, although the latter would cast a shadow on any surface behind it. For this reason a medium thickness of line seems best suited to the position. As to the lining-in of the plan P of the cube, it will at once be seen from its position that only one edge, or that between points *f* and *b*, will cast a shadow, and it must therefore be made a thick line. The other three —two in the plane of the light and one having light falling direct upon it—must be fine lines.

38. To prevent the misreading by the student of some engravings of mechanical subjects in outline still to be met with, it is as well here to observe that in England some draughtsmen and engravers line-in both plans and elevations of objects as if they were situated *in the same plane*, and cast their shadows in the same directions; but that this is contrary to the principles upon which their projections were obtained may at once be seen from a study of the shadows cast by the cube in plan and elevation in Fig. 133. As the planes of projection are at right angles to each other, it would be impossible for the shadows to fall in the same direction in *both* planes, unless the light fell upon the objects represented in two totally different directions at one and the same time, which would be absurd. In all representations of objects, therefore, throughout this work, where any difference in the lining is shown, we shall recognize the existence of the two planes of projection, in their *proper relative positions*, and apply the rules for such lining in accordance with the principles before explained.

As the study of the projection of solids having *curved* surfaces will be the next division of our subject that will occupy attention, a few words only are here necessary with reference to the lining-in of such objects in ink. As a general rule, only *one* kind of line—the *fine* line— is applicable to the outline representation of the curved surface of the solids of this class—the cylinder, the cone, and the sphere—with which we shall deal, consequent on the fact that the particular part of the curved surface of each which would cast a shadow on any adjacent surface, falls *within* the bounding lines of its representation, and cannot be expressed by a line, as in the case of bodies having plane surfaces. We shall, however, before completing our explanation of the projection of such solids, give all necessary information as to inking them in.

39. Before leaving this part of our subject a few words are necessary for the guidance of the student in reference to the mixing and usage of the ink with which he will line-in his drawings. Although

MECHANICAL AND ENGINEERING DRAWING   83

liquid ink for this purpose is now to be purchased, it is advisable for the student to know how to make suitable ink for himself. He will not, of course, without some considerable practice, succeed in making really serviceable ink, but such as he will be able to produce will sufficiently serve his present purpose.

To prepare the ink for use, let the student provide himself with a medium-sized colour saucer—sold in nests of half-a-dozen by any artists' colourman—and having procured a stick of good India-ink and a camel's-hair water-brush, proceed as follows :—Charge the brush with clean water, and by its means put as much of it in the saucer as would half-fill an egg-spoon. With the stick of ink held vertically, and with one end in the saucer, rub it in a circular direction, applying a gentle pressure at first, until the water has attained a dark brown colour and begun to thicken ; continue rubbing, gradually adding water in drops from the water-brush, until a sufficient quantity of liquid ink is made. To test its blackness and consistency, blow steadily into the saucer : if the ink is a good black all through, the bottom of the saucer will not be seen ; but if too thin, it will come in sight, and the rubbing must be continued. For the ink to be in working order it should be neither too thin nor too thick. To bring it to a proper consistency, about a quarter of an hour's rubbing will be required. If it is found at any time to have thickened by evaporation of the water in it, add more in drops by the use of the water-brush, keeping it well stirred while doing so.

Some writers on this subject advise the student to always use *fresh-rubbed* ink, taking care to wash out the saucer before making it. This is by no means the practice in many of the largest drawing offices in the kingdom. On the contrary, good ink is never wasted, but is improved by being worked up afresh. The writer's practice for years has been to keep by him in his instrument-box a home-made india-rubber pestle, similar in shape to a diminutive Indian club. With the use of this and a few drops of clean water added as required in the ink-saucer, he is able to produce an ink of equal density and colour throughout, and which does not contain a particle of anything held in suspension in it that would be liable to clog the drawing-pen.

Special unglazed saucers are now to be purchased for the rapid making of India-ink, but these the writer finds very extravagant substitutes for patience and the expenditure of a few minutes of time. A more suitable saucer would be one made of glass of the same shape as now used, but with its inner surface finely ground. Through want of patience on the part of the draughtsman, and the application of too much pressure while rubbing, the ink is liable to break away from the stick in small lumps, which are never thoroughly reduced to the liquid state, but cause constant annoyance by clogging the pen. With a properly-ground glass saucer this tendency of the ink-stick to fray away in particles would be avoided. One great mistake made by many draughtsmen is the practice of *dipping* the rubbing end of the ink-stick in water before starting to make their ink. By so doing it becomes soddened, and when dry will be found to be cracked all over at that end, and ready to break away in pieces on the least pressure being applied. The rubbing end of the ink-stick should always be wiped dry

MECHANICAL AND ENGINEERING DRAWING

directly it is done with, if it is wished to keep it in good order for use ; and the ink saucer should only be uncovered while filling the pen. This latter operation is best effected by capillary attraction, and not by the use of a brush. When more ink is required, first pass a piece of soft wash-leather between the nibs of the pen, to remove any deposit, and before dipping in the ink draw it through the lips, which will give sufficient moisture to cause the ink to flow freely between its nibs, and so fill it. The use of a brush is objectionable, on the ground of its being liable to pass into the pen any sediment that may have collected in the ink-saucer.

With reference to the drawing-pen and how to use it—the compass or bow pens will not at present be required—the student will remember the explanation of its construction given at the commencement of this work, and from that will understand that in using it, it must be held nearly upright, keeping its *strong* nib in contact—without pressure —with the edge of the tee- or set-square ; that all horizontal lines must be drawn from left to right, and vertical ones upwards, or from bottom to top of the board. The drawing-pen when purchased should be *set* for use, and its satisfactory manipulation will soon be acquired if due care is taken not to damage its points.

CHAPTER XII

THE PROJECTION OF CURVED LINES

40. UP to the present stage in our subject we have purposely refrained from dealing with figures having *curved* bounding lines, or solids with *curved* surfaces, knowing how important it is for the student to thoroughly understand the projection of straight-lined plane figures and plane-surfaced solids, before attempting anything in which a curved line or surface occurs. Assuming that he has mastered all that has now been explained, he will be the better able to apply himself to that part of the subject to which we next proceed—viz., the projection of solids having "curved" surfaces. As this, however, involves a knowledge of the projection of *curved* lines, this must first be acquired.

From our definition—Chapter III.—of a curved line—viz., that it is the path of a *point* which is continually changing its direction—it will at once be manifest that its projection cannot be obtained in the same way as that of a straight line, in that all that is here required is to find the projections of its ends and join them by a right line. As the generating point of a curved line is moving in a different direction at all parts of its path, it follows that the only way in which its projection can be obtained is by getting the projection of that point in several of its positions, and joining them by a line passing through them, which will be the projection sought To obtain this, however, it is first necessary to have a perfect knowledge of the original curved line and its varying direction, as it may be a line lying wholly in a plane, or, on the other hand, one of such a character that no part of it would coincide with or lie in a plane. For a first example in curved line projection we take as our problem—

Problem 44 (Fig. 138).—*Given the front elevation AB of a curved line in a plane parallel to the VP, it is required to find its side elevation and plan, or views of it when looked at in the direction of the arrows.*

Here the line being a simple curve, it is evident, on looking in the direction of the arrow on the left, that its projection will be a straight line *ab*, No. 1, as all the points in the curve are in one and the same

FIRST PRINCIPLES OF

plane. For the same reason, the plan of this curve will be the straight line $a'b'$, No. 2, obtained by letting fall projectors from its two ends AB into the HP, and joining them by a right line.

Now, let the plan of this curved line—or $a'b'$, No. 2—be assumed to make an angle of 40° with the VP, and its front and side elevation be required. To obtain these, we must know the position of some points in the curve with respect to a fixed *datum* line, from which measurements may be taken. Let the points in the curve of which $a'b'$ is the plan be A, 1, 2, 3, B, in the elevation, and let the line CB, drawn through B at right angles to the IL, be the datum line, which is assumed to be in the same plane as AB. Through A, 1, 2, 3, B, draw lines parallel to the IL, cutting CB in points C, 1', 2', 3', B. Now, the line $a''b'$ in the HP is the plan of the curved line AB in elevation, moved through an angle of 45° with the VP, the point a'' being nearest the eye, and b' farthest from it. To find the plans of the intermediate points 1, 2, 3, in the curve AB, set off from b' in the plan the distances CA, 1 1', 2 2', 3 3', in elevation in the points a'', x, y, z. Through these draw projectors into the VP, cutting those from A, 1, 2, 3, in c, d, f, g. Then a line drawn through these points to B will be the required elevation of the curve.

To find its side elevation, take a new datum line C'B', No. 3, parallel to CB, and produce the projectors drawn through A, 1, 2, 3, B, in the original curve line AB, No. 1, to cut it in points C', 4, 5, 6, B'; at each of these set off in points c', d', f', g', the distances—measured from the datum line—that their corresponding points a'', x, y, z, in the plan No. 2 are from $a'b'$, measured on the projectors drawn through them; then a line drawn through the points so obtained, as shown in No. 3, will be the side elevation of this curved line in its new position.

If the given line of which the projections are required is one of double curvature, but still in one plane, its projections are obtained in the same way as those of a simple plane curve. As an example we will take a very common double-curved line known as the *ogee* moulding; and let the problem be—

Problem 45 (Fig. 139).—*Given the front elevation of a line of double curvature AB, No. 1, to find its plan and side elevation, the plane of the curve being parallel to the VP.*

In this case, as every point in the curve is at an equal distance from the VP, its plan will be a straight line $a'b'$, No. 2, parallel to the IL, and its projected length will equal the distance between projectors let fall from its ends A and B into the HP. Its side elevation will also be a straight line ab, at right angles to the IL, for as every point in it is equi-distant from the VP, it will coincide with or lie in a plane which we are told in the problem is parallel to the VP, and therefore perpendicular to the HP.

To find the elevation of the line AB when the plane in which it lies makes any given angle with the VP, proceed as follows :—

Choose any convenient points—other than its two ends—in the curve as 1, 2, 3, 4, No. 1. Through these let fall projectors into the

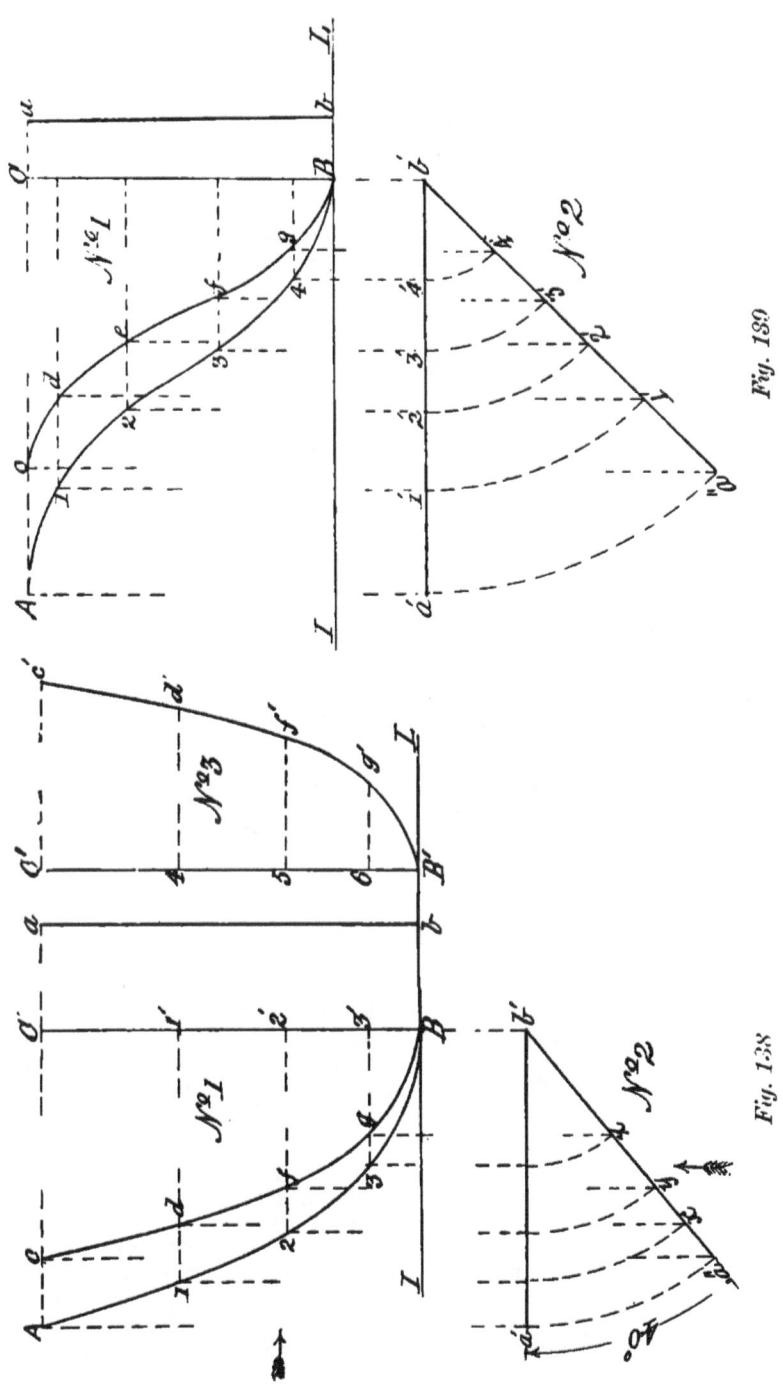

Fig. 138

Fig. 139

88   FIRST PRINCIPLES OF

HP to cut *a'b'*, No. 2, the plan of AB in the points 1', 2', 3', 4'.   At *b'*, in this plan, draw a line making with *a'b'* an angle equal to that—say 45°—the plane containing the curve line is to make with the VP Transfer—by arcs struck from *b'*—the points 1', 2', 3', 4' in *a'b'* to *a"b'*, and through these draw projectors into the VP, cutting those drawn through A, 1, 2, 3, 4, No. 1, to the datum line CB in *c, d, e, f, g ;* a curve drawn through these points, as shown, will be the elevation required.   In this way the elevation of any curved line lying in a plane making any angle with the VP, may be found, for the line, or the plane in which it is supposed to lie, may be assumed to rotate round the datum line as an axis, carrying with it all its points, the position of which with respect to the VP may always be easily determined, and from them the projections of the lines in which they lie may be found.

41. To find the projections of a compound curved line, or one no part of which would coincide with or lie in a plane, the process is somewhat more difficult, as its appearance in two directions at right angles to each other must be known, or given, before a third view of it can be obtained.   As an example—

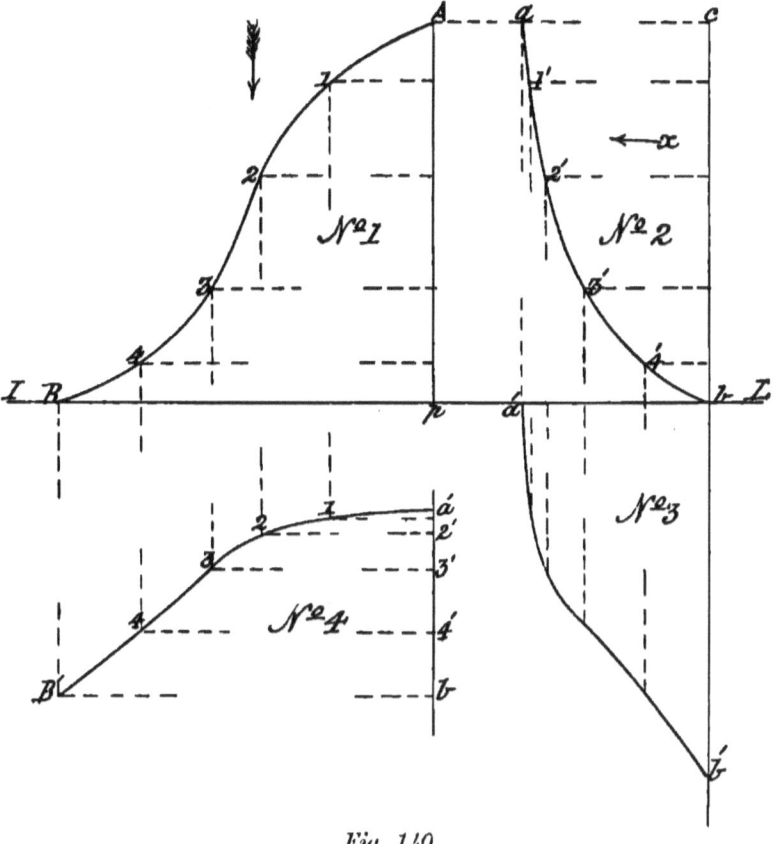

Fig. 140

MECHANICAL AND ENGINEERING DRAWING 89

Let AB, No. 1 (Fig. 140), be the elevation of such a curved line, and *ab*, No. 2, another elevation of the same line, at right angles to the first. It is evident that such a line could not coincide with a plane, yet if the relation of any points in it with the planes of its projection be known, or can be determined, then the plan or horizontal projection of the line can be found. Let, then, the line *ab*, No. 2, be the front elevation of the line, the VP being behind it, and AB, No. 1, its side elevation, looking in the direction of the arrow *x*. Then the end A or *a* of the line is nearest the VP, or farthest from the eye; and the end B or *b* the converse. To prove this, through A, No. 1, and *b*, No. 2, draw the lines A*p*, *cb*, perpendicular to the IL; then A*p* will represent an end elevation or edge view of the VP—looking in the direction of the arrow *x*—and *cb* that of a plane at right angles to the VP. With the assistance of these as *datum* planes we can determine the distance of any point in the line AB from each of them, and thus obtain its horizontal projection. A line drawn through a series of points thus found will be the plan of the curved line sought.

To find it, choose a suitable number of points in the given line AB, No. 1, say at 1, 2, 3, 4; project these over on to *ab*, No. 2, and produce the projectors to meet the datum plane *cb*, as shown; the exact position *in plan* of any of these points can now be fixed. From *a*, 1', 2', 3', 4', *b*, No. 2, let fall projectors into the lower plane, or HP, and on these, in No. 3, set off from the IL the distances that the corresponding points in AB, No. 1, are from A*p*; through the points thus found draw the line *a' b'*, and it will be the plan of the given compound curved line sought. By letting fall similar projectors from the points in AB, No. 1, and setting off on them the distances that their corresponding points in *ab*, No. 2, are from the datum plane *cb*, a similar plan of the compound curved line will be obtained, as shown at *a'* B', No. 4, to that found in No. 3, but at right angles to it, as AB, No. 1, is at right angles to *ab*, No. 2.

From the foregoing explanation the student should be enabled to apply the principles involved to the projection of any single curved line, whether simple or compound; but to show the actual appearances the same curved line will assume in projection when occurring in different positions, we give one problem in this connection before passing on to the projection of curve-bounded figures.

By way of making the problem more interesting, and of practical application, we will take an ordinary curve—such as is given to the foot of a square or six-sided column—combined with a straight line, and let the problem be—

Problem 46.—*Given the elevation of a combined curved and right line parallel to the VP, to find its plan and elevation when it makes any given angle with the VP or HP.*

Let ABC, No. 1 (Fig. 140a), be the elevation of the combined line. Being parallel to the VP, every part of it is equi-distant from it. Its plan, in this position, will be the straight line *a'b'*, No. 2, parallel to the IL, its length being determined by projectors from A and B in the elevation. Produce BC, No. 1—the straight part of the line—vertically to c', and let c'C be assumed as an axis round which the curved line can be turned through any given angle; and also as a datum line from which measurements may be taken. In AB—the curved part of the line—take any convenient points 1, 2, 3, 4; and through them and A draw projectors parallel to the IL, to meet the axis, or datum line, in c', $1'$, $2'$, $3'$, $4'$. Now, suppose ABC to rotate on the axis line c'BC, from its normal position—parallel to the VP—through any given angle; then its plan at that angle will still be a straight line, but it will be at an angle with the IL—or with its plan when in its first position—equal to that it has been turned through. Let this angle be one of 60°. Then to find its elevation in this new position, proceed as directed in solving Problem 45 by first projecting over on to ab in plan the points in the curve, transferring them by arcs to $a''b'$, and from thence by projectors into the upper plane, cutting those drawn through A, 1, 2, 3, 4, B, in the points shown; then a line through a' and these points to B will be the elevation of the curve in its new position.

Next, let this newly-found elevation of the combined line be inclined to the HP in such a way that the axial line—and with it the part BC of the combined line—makes an angle of 45° with the HP or IL; and let its plan when in that position be required.

To put on the paper the new elevation, No. 3, in the required position, draw in the axial line at the required angle, and to it transfer the points in the same line, No. 1; through these draw ordinates as shown, and on them set off in points a', 1, 2, 3, 4, the corresponding lengths of the same lines in No. 1, and draw in the curve. To obtain its plan in this position, let fall projectors from the several points in it in the elevation, and through the corresponding points in the plan, No. 2, draw projectors parallel to the IL, cutting those let fall from No. 3; a line drawn through the points of intersection of these projectors will give a' BC, No. 4, the plan of the line required.

In the foregoing, Nos. 1 to 4, the axial line about which the combined line has been supposed to rotate has in each case been assumed to be parallel to the VP. Let it now be supposed, while still inclined to the HP at the same angle as before, to be also inclined to the VP at an angle say, of 45°, and let an elevation of the combined line in this position be required.

To obtain this, we must first show in on the paper, the plan of the combined line in its new position with respect to the VP or IL, before we can proceed with its elevation. This means that the last-obtained plan of the combined line must be moved from its present position—viz., that of having the assumed axial line c' C parallel to the VP—to one in which that line will make an angle of 45° with the VP or IL; and is tantamount to swinging the line on its end C, as a pivot, through an arc of 45° in such a way that each point in it while being swung is kept at the same height—in its plane of rotation—above the HP. This new position of the last-found plan of the line is therefore correctly shown, by drawing a line—the assumed axial one—at an angle of 45° with the IL, and transferring to it, as was done in the case of

MECHANICAL AND ENGINEERING DRAWING

91

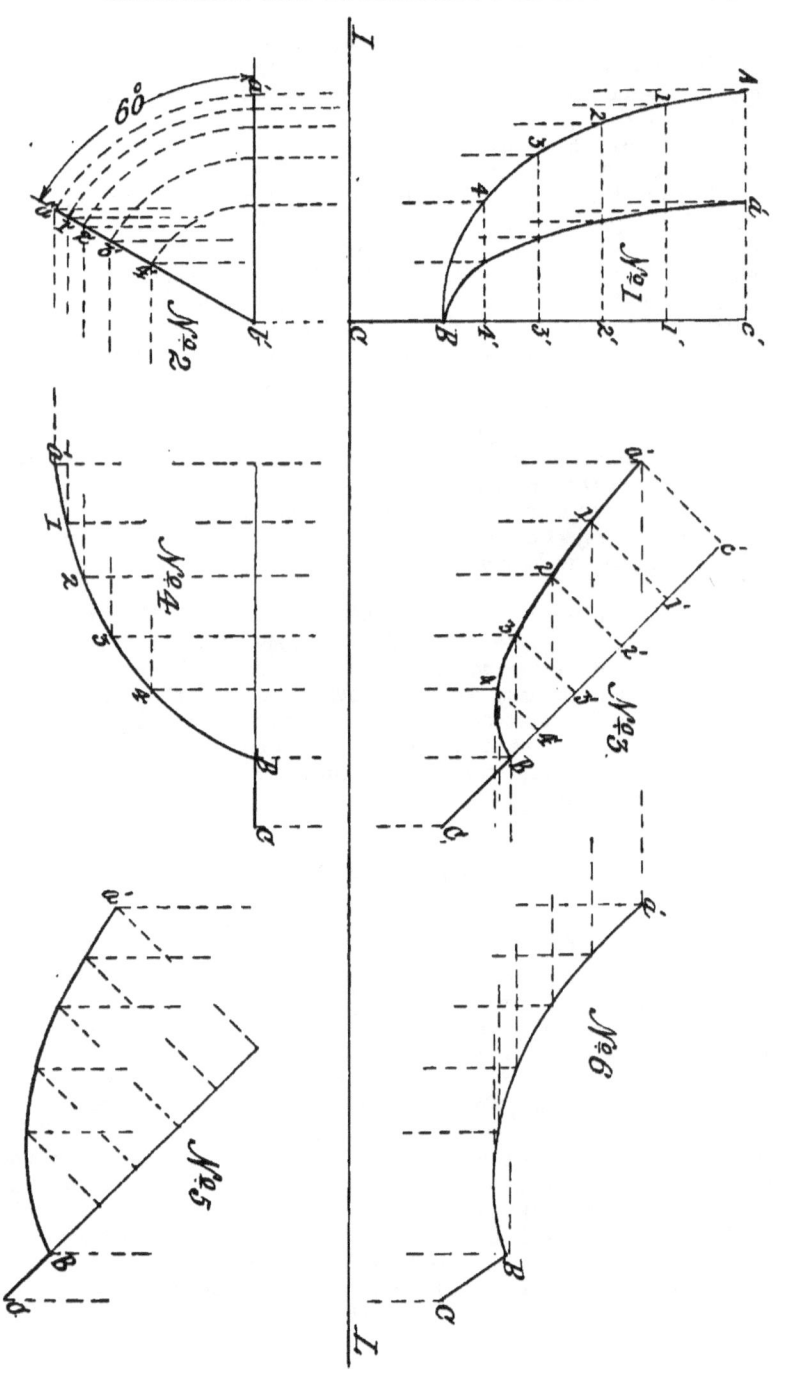

92   FIRST PRINCIPLES OF

No. 3, the various points, ordinates, etc., of No. 4, as given in No. 5. Its elevation is then obtained by producing the projectors drawn through the points a', 1, 2, 3, 4, BC, No. 3, and intersecting them by those drawn through the corresponding points in the plan No. 5. A line drawn through the points where these projectors intersect will give the line a'BC, No. 6, which is the required elevation.

42. To familiarize himself with the principles involved in this very interesting part of our subject—viz., the projection of curved lines in *any* position—the student should accustom himself to make wire templets, of copper or lead, bent to the exact shape of the lines given for projection. Such models, when posed before him in the positions stated in the problems, would materially assist him in obtaining the correct projections required, as he would be able to follow the movements of any points in them in the different positions they assume. Practice with such models is all the more advisable in view of the more difficult problems that will have to be mastered in the projection of solids having curved surfaces, when inclined, or otherwise, to the planes of their projection.

After the foregoing full explanation of the application of the principles of projection to curved lines, the projection of curved-bounded figures, to which we now pass, will present little or no difficulties.

As the simplest possible curve-bounded figure is the circle, its projections claim first attention. We have defined such a figure as the path described by a *moving* point that is always at the same distance from a fixed point, around which it moves. When such a figure is cut out of any material substance, such as a piece of metal plate, or card, it is called a *disc*. Looked at in various directions, it assumes different forms. The exact form taken is obtained by projection. Our first problem in connection with it is—

Problem 47.—*Given the plan of a circular plate, to find its front and side elevation.*

Let the straight line AB, Fig. 141, No. 1, be the given plan. Now as the edge view of a disc, looked at from above, is a straight line, we see that if AB is such a view, the plate must be perpendicular to the plane of its projection, or the HP, and being parallel to the IL every part of its surface must be parallel to the VP; therefore its elevation will be a circle and have its bounding-curved line everywhere equidistant from its centre. To find this bisect AB in C, and through C draw a projector into the upper plane at right angles to the IL. Take any convenient point in this projector at a greater distance from the IL than half AB—say a'—and through a' draw a line parallel to the IL. With a' as a centre and AC, No. 1, as a radius, describe the circle *acbd*, No. 2, which will be the front elevation of the circular plate. If correctly drawn, vertical projectors through A and B in the plan should cut the diametral line drawn through the centre a' of the circle in the points a and b in that line. Then as the surface of the plate is parallel

MECHANICAL AND ENGINEERING DRAWING   93

to the VP, the side elevation of it will be the straight line $c'd'$, No. 3, parallel to cd, No. 2, obtained by projectors from c and d, as shown.

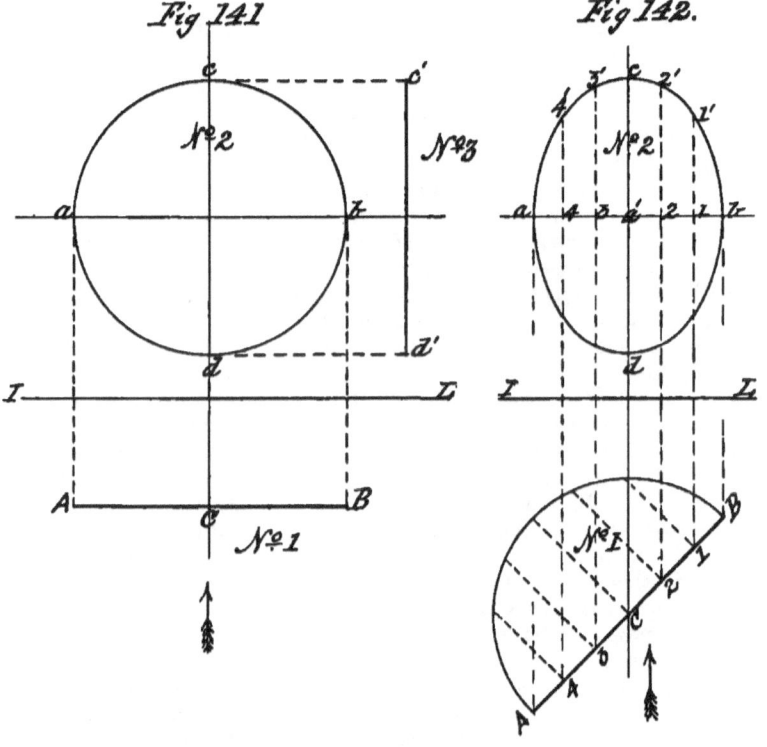

Next, let the line AB, Fig. 142, No. 1, be the given plan of the circular plate, and its elevation be required.

Here the plate being still represented in plan by a line, is perpendicular—as in the previous case—to the HP, but being inclined to the VP its elevation becomes an *ellipse*, or a curve-bounded figure in which points in its bounding line are not all the same distance from its centre. To find the elevation, bisect AB as before in C, and through C draw Cc, at right angles to the IL. Choose any point in Cc, as a', and through it draw a line parallel to the IL. Set off in c and d from a' on the projector through C, a distance equal to AC or CB, and through A and B draw projectors into the VP, cutting the line through a' in a, b; then $acbd$, No. 2, will be four points in the ellipse, the line between c, d being its *major* axis, and that between a, b its *minor* axis. Divide AB in the plan into any number of equal parts, say six, in the points 1, 2, C, 3, 4 ; through these at right angles to AB draw ordinates to the semi-circle. Project points 1, 2, 3, 4, in AB on to the minor axis ab in the elevation, and through them draw lines parallel to cd. Set off on these on either side of *ab*, the length of the corresponding ordinates to the semi-circle in No. 1, and through the points thus obtained draw the curve of the ellipse as shown in No. 2, which will be the required elevation of the circular plate in its new position.

43. For the better comprehension of this last problem by the student, before passing to the next a few words in further explanation of its solution is desirable. It has been shown in previous problems in the projection of curved lines that if the *position* of points in the curve can be determined the problem is virtually solved. In the case of the circle in Fig. 141, it would have been quite possible to have first found any number of points in the bounding line and then to have joined them by a line passed through them, and thus have produced the circle, but the use of the compass obviated the necessity. In the problem before us, the necessity exists for finding exactly several points in the required elevation of the object in the readiest possible way. The original object being a circle, a regular figure—as distinguished from an irregular one—whose surface is readily divisible into parts, it is apparent that by turning down one half of that surface on the line AB in No. 1 as a diameter, and drawing lines upon it at right angles to AB, through any points in the semi-circle, we can at once find the actual length of those lines and transfer them to their vertical projections in No. 2. The solution of this problem may also be reasoned out in another way. The original line AB, No. 1, is in fact—as was shown in the previous problem—a curved line, although presented to the eye as a straight one, every point in it, on either side of that at C, being farther from the eye the farther it is to the right or left of C. As the actual length of a line drawn across the circle at C, through its centre to the opposite edge, is known to be equal to AB, or twice Cc, so can the distance across the circle through any other point in AB be ascertained in the same way. For, as the ordinate Cc is measured at right angles to AB, so any other, as 2, 2′, is equal to half the distance from the point 2′ across the circle to the corresponding opposite point in its edge. With this explanation the following problem should present no difficulty in its solution.

Problem 48 (Fig. 143).—*Given the plan of a circular plate as a circle ; to find its projections in the VP and HP when making a given angle with either plane.*

Let the circle ABCD, No. 1, be the given plan, the diametral line AB being parallel to the IL, and the similar line CD perpendicular to it. Bisect each quarter of the circle in the points 1, 2, 3, 4. As shown, the surfaces of the plate are evidently parallel to the HP. Let it now be inclined to that plane at an angle of 60°, keeping the diametral line AB parallel to the HP, and the point C uppermost, or nearest to the eye. To obtain its elevation when in that position, an edge view of it—looking in the direction of the arrow *x*—must first be given, which will be a line equal in length to CD, inclined to the IL at the required angle. Draw in this line CD as in No. 2, and bisect it in the point B, which will be the end of the diametral line AB, nearest

the eye in side elevation, A being the point beyond it across the
circle. Through B, No. 2, draw a projector parallel to the IL, and intersect it in points a, b (No. 3) by projectors from A and B in No. 1.
Transfer points 2 and 4 in No. 1 to 2 and 4 in No. 2, and through
them draw projectors parallel to that through B; then from 3, 4, in
No. 1, intersect by projectors—which will pass through 1, 2—those
drawn from 2, 4, No. 2, in 1', 2' 3', 4', No. 3. Through the eight
points thus found, draw the closed curved-bounding line shown, and
the figure or ellipse will be the front elevation of the circle No. 1 in
plan when inclined at an angle of 60° to the HP, and 30° to the VP.

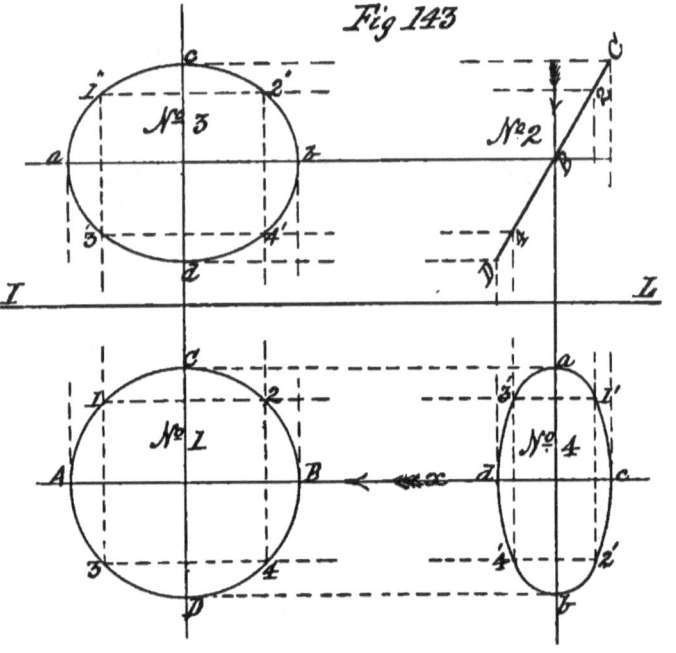

Fig 143

To find the plan of the plate in this position, let fall projectors
from the points in No. 2 into the HP, and their intersection by those
drawn through C, 2, B, 4, D, No. 1, will give points through which
the ellipse No. 4—which is the plan of the circular plate—is to be
drawn. It is here seen that the form assumed by a circular plate
—or a circle—entirely depends upon the direction in which it is
viewed, alternating, as it may be made to do, from that of being a true
circle—or any form of ellipse—to that of a perfectly straight line.
The student will note what may appear to be a discrepancy in the
lettering of the last-obtained projection of the original object, No. 1,
but he will find that it is correct when he remembers that the view of
it given in No. 2 is that of No. 1 looked at in the direction of the
arrow x, and therefore turned through an angle of 90°. If a disc,

96   FIRST PRINCIPLES OF

similar in size and similarly figured, be cut out in card and laid upon
No. 1 with a pin through the centres, it will be found, on turning the
card disc through an angle of 90°, that all the points on it will have
shifted through that angle, bringing A and B where C and D now are
thus showing the projection No. 4 to be correct.

44. Before leaving this part of our subject, as *elliptical* figures
will be very frequently required, we here give a simple method of
drawing such a figure, which is sufficiently near the true form for all
the practical purposes of the draughtsman, and which obviates the

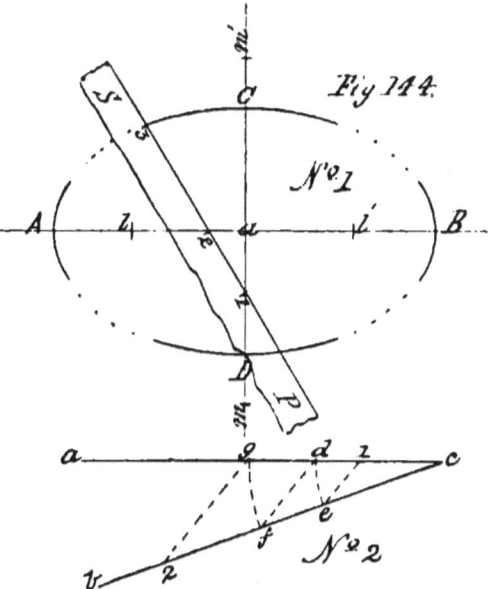

Fig 144.

necessity of finding all the points in the curve. Premising that the
student understands that an ellipse differs from the circle in having its
diameters of unequal length, and that the sum of the distance of any
point in its curved-bounding line, from two fixed points—called its
foci—in its longest diameter is always the same, then draw an ellipse
by the method above referred to :—

Let the line AB, No. 1 (Fig. 144), be its major or longest diameter
or axis, and CD, drawn at the mid-length of and at right angles to
AB, its minor or shortest diameter or axis. Anywhere on the sheet
of drawing paper draw two lines, as *ac*, *bc*, No. 2 making an angle
with each other—say 30°—intersecting at *c*. On *a c* set off with the
compasses from *c*, half the length of AB, No. 1, and describe the arc
de, and on *bc* with half CD as radius, and from the same centre
describe the arc *fg*, join *df* by a right line, and parallel to it draw
lines through *e* and *g*, cutting *ac* in point 1, and *bc* in point 2. Then
on the major axis AB, No. 1, set off from A and B the distance *cl*

MECHANICAL AND ENGINEERING DRAWING   97

No. 2, in the points ll', and with it as a radius from ll' as centres, draw arcs of about 30° through A and B. On the minor axis CD set off from C and D, the distance $c2$, No. 2, in the points mm', and with it as a radius and mm' as centres draw through C and D arcs similar to those drawn through A and B. For intermediate points take a straight-edged slip of paper SP and mark on its *edge* with a fine-pointed pencil in points 1 and 3, half the longer axis aA, or aB, No. 1, and in point 2—measured from point 1—half the shorter axis aC, or aD. If, then, this slip of paper be laid on the figure No. 1 and moved over it in such a way that point 2 in its edge is always on the major axis, while point 1 is on the minor, point 3 will describe, if carried through a whole revolution, a perfect ellipse. As, however, only a few points are required between the ends of the arcs already drawn, these can be marked off from the edge of the slip of paper by moving it in the desired consecutive positions.

CHAPTER XIII

THE PROJECTION OF SOLIDS WITH CURVED SURFACES

45. THE particular solids with which we have here to deal are those known as "solids of revolution," which are assumed to be "generated," or produced, by the revolution of certain geometrical figures round one of their bounding lines as an axis; the other lines in their circling round this axis enclosing a space—assumed to be material—of a certain form, dependent upon that of the generating figure. A "solid of revolution" therefore implies an object having curved-bounding surfaces. Of such solids there are three primary ones of which machine details are made up—viz., the *cylinder*, the *cone*, and the *sphere*—the subsidiary ones being generated by figures which are portions of plane sections of the first two primary ones.

A *Cylinder* is a solid generated by the revolution of a rectangle about one of its sides as an axis. If ABCD, Fig. 145, be the rectangle, then if it be caused to revolve round AB as an axis, making one complete circuit, the sides AD, DC, CB will generate the solid called a "cylinder," having a height, or length, equal to DC, a diameter equal to twice AD, and its ends E, F, circular flat surfaces, at right angles to its axis AB.

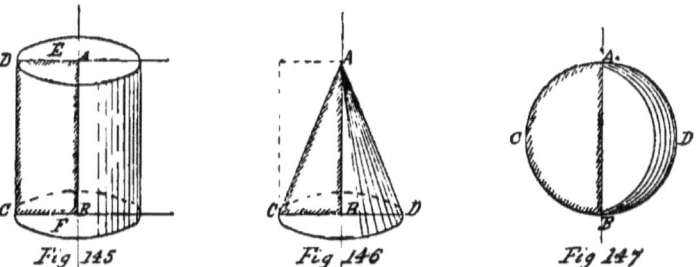

Fig 145   Fig 146   Fig 147

A *Cone.*—If one half of the rectangle—divided diagonally—Fig. 146, be removed, leaving the right-angled triangle ABC remaining; then, if this triangle be revolved on AB as an axis, as before, the solid generated will be a "cone," as in Fig. 146, its base DC being a circular flat surface at right angles to its axis AB.

98

MECHANICAL AND ENGINEERING DRAWING   99

A *Sphere.*—Again, if the two lines AC, BC of Fig. 146 be removed and a semi-circle be described on AB as a diameter, as in figure 147, then on revolving this figure AB as an axis, as before, the solid generated will be a "sphere," as in Fig. 147.

As the subsidiary solids before referred to are generated by parts of certain sections of the cylinder and cone, their definition is deferred until those sections have been found by projection. Taking the "cylinder" as the first curved-surfaced solid as our object, the problem is —

Problem 49 (Fig. 148).—*Given the plan of a cylinder with its axis perpendicular to the HP, to find its elevation when its length is twice its diameter.*

Let the circle AB, No. 1, be the given plan; then its centre *a* will be the plan of the axis of the cylinder. Find by projection the elevation of this axis *a′ a*. Assume the cylinder to be standing with one end on the HP; then, as its *ends* are in the same relative position as the *sides* of the rectangle which generated them—viz., parallel to each other—and one of them is on the HP, set off on the axis, from the IL, the length of the cylinder in the point *a′*, and through it draw a line parallel to the IL. Now, in looking at the cylinder in the direction of the arrow in the plan No. 1, the visual rays will impinge upon its surface from A to B; at A and B the rays will be tangential, and being at the same time perpendicular to the plane of projection, or the VP, they will strike both sides of the cylinder in lines drawn through A and B on its surface, perpendicular to the HP. Therefore through A and B, No. 1, draw the lines AC, BD, No. 2, and the required elevation is obtained.

Now, let the cylinder be inclined to the HP, at an angle of 45°—its axis being still parallel to the VP—and its plan when in that position be required.

First draw in the elevation of the cylinder in the given position, as in No. 3. Its ends AB and CD are now inclined to the IL or HP, and will in plan become ellipses—as explained in Problem 40—because they are *circular*, but inclined to the plane—the HP—on which their projections are required. Now, in viewing the cylinder in this position from above, or in the direction of the arrow, only one of its ends, AB, will be seen, the other, CD, being by its inclination invisible. To find its plan when so inclined, first draw in the plan of its axis *a′a*; on it obtain by projection the plan of the end AB, No. 3, as shown by the ellipse A1, B2; through points 1 and 2, in this ellipse, draw lines parallel to the IL, which will be plans of the sides of the cylinder seen from above, or along the line *a′a*, No. 3; then find by projection the plan of that part of the bottom edge of the cylinder that will be seen, and the required plan will be obtained.

Again, let the cylinder, as in No. 5, have its axis inclined to the VP, at an angle of 39°, but parallel to the HP, and its elevation be required.

To obtain this, proceed as in the last case, by getting first the

100   FIRST PRINCIPLES OF

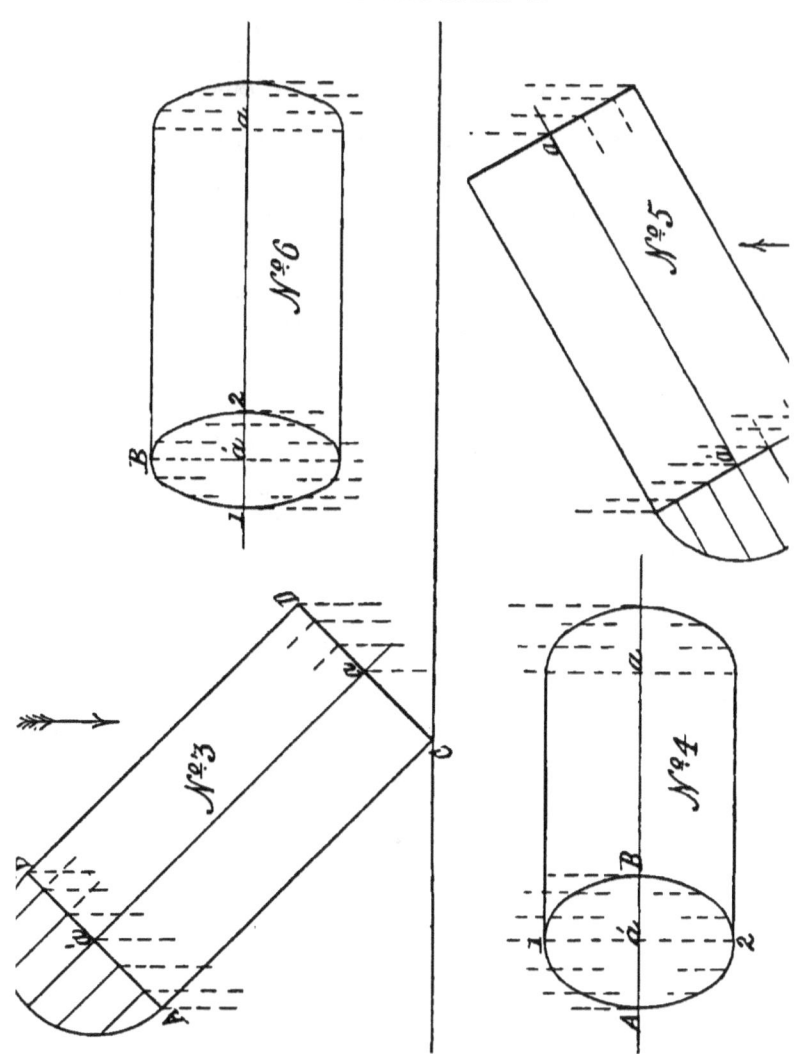

MECHANICAL AND ENGINEERING DRAWING   101

elevation of the axis, and then finding by projection the two ends of
the cylinder upon it, and drawing in its sides as shown in No. 6.

For the projections of the " cone " the first problem is —

Problem 50 (Fig. 149).—*Given the plan of a cone, having its axis
perpendicular to the HP, and an altitude equal to twice the diameter of its base, to find its elevation.*

Let the circle No. 1, Fig. 149, be the given plan of the cone ; its
centre *a* as in the cylinder will be the plan of its axis. Find by projection the elevation of this axis, and on it set off from the IL the
altitude of the cone in point *a'*. Through *a*, the centre of the circle in
No. 1, draw AB parallel to the IL. Find by projection the elevation
of points A and B—in the base of the cone—on the IL, and join these
by right lines with point *a'* on the axis *a' a*. The triangle A *a'* B will
be the required elevation.

*Next, let the base of the cone be inclined to the HP at an angle of 45°,
keeping its axis parallel to the VP, and its plan be required.*

Here, an elevation of the cone at the given angle must first be
drawn in, as in No. 3 ; then find the plan of its axis, which, being
parallel to the VP, will be a line parallel to the IL, as in No. 4.
Obtain by projection the plan of the base of the cone, which is an
ellipse, its major axis being CD and its minor AB, No. 4 ; then find
the plan of the apex *a'*, No. 3, of the cone in *a'*, No. 4, and join it by
right lines to points C and D ; the required plan of the cone in the
position stated will thus have been obtained.

*Again, let the axis of the cone be parallel to the HP, but inclined to the
VP, at an angle of 45° ; and its elevation be required.*

To obtain this, draw-in the plan of the cone in the position
stated, then find the elevation of its axis, and on this, by projection
from the plan, draw-in the base, which will be an ellipse ; join the two
ends of its major axis with the projected apex as shown in No. 6, and
it will be the required elevation.

As a "sphere," when looked at from any direction, is a circle,
when projected, we shall defer any problems in connection with it
until we come to consider the projections of its sections—which will
follow those of the cylinder and cone, to which we next give attention.

102   MECHANICAL AND ENGINEERING DRAWING

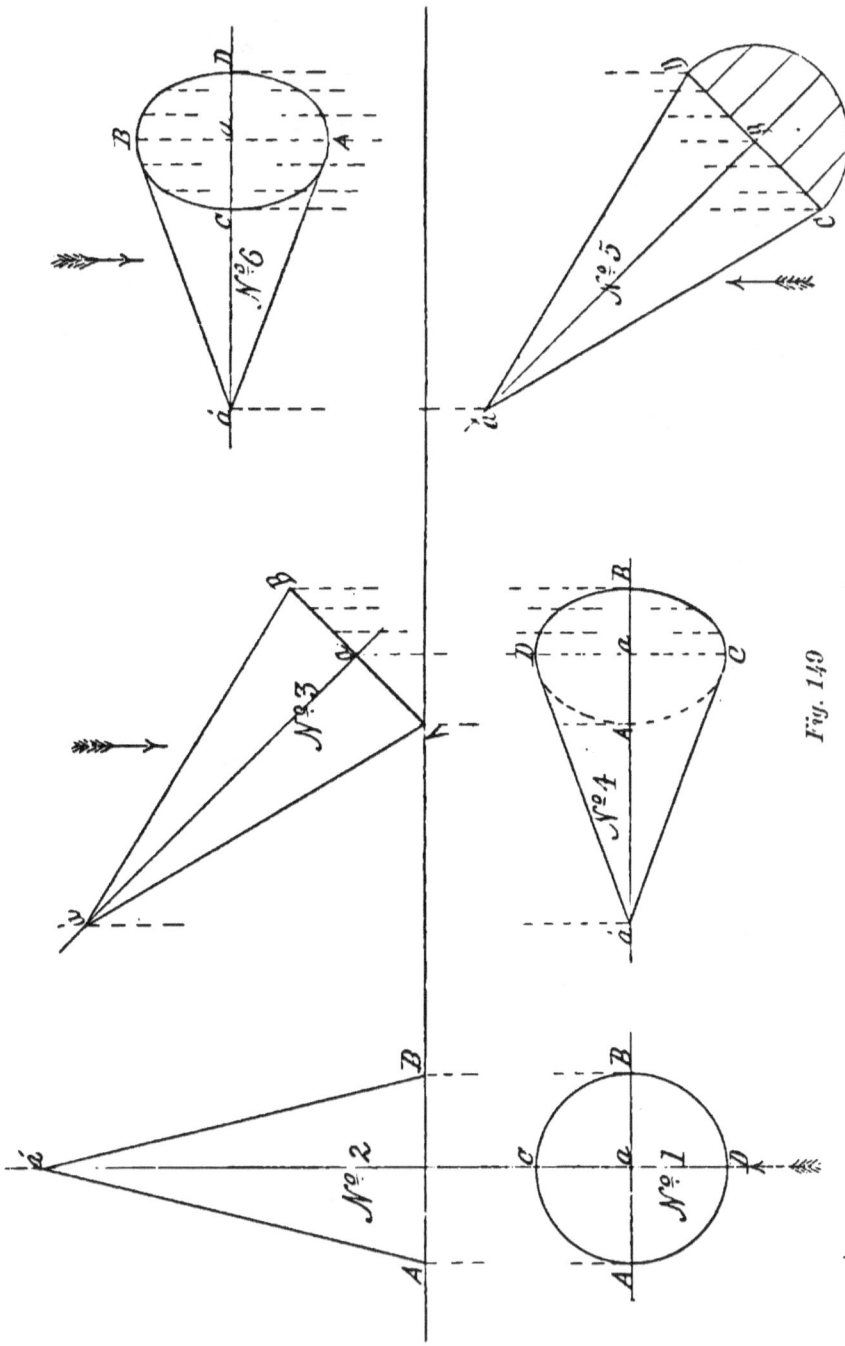

Fig. 149

CHAPTER XIV

THE PROJECTION OF THE SECTIONS OF A CYLINDER

46. FROM the definition previously given of a "cylinder," it follows that any and every point in its curved surface is at the same distance from its axis; every plane section of it taken parallel to the axis is a rectangle, while every section at right angles to its axis—or parallel to its ends—is a circle. These are self-evident truths, and require no graphic illustration to prove them. It is then with its sections *not* taken in either of the directions mentioned that we shall now deal. Our first problem in connection with them is—

Problem 51 (Fig. 150).—*Given the elevation of a cylinder, cut by a plane making an angle with its axis; to find the sectional elevation.*

Let the cutting plane KP, No. 1, make an angle of 45° with the HP, and be perpendicular to the VP, and the view required be that in the direction of the arrow. To solve this problem, we must know the distance between any point in the line of section on the *front* side of the cylinder, and its corresponding point on the side of the cylinder *nearest* the VP. To determine this, first find the plan of the cylinder No. 2, and project over in No. 3 the elevation of the lower part of the cylinder not affected by the cutting plane. Then to find the contour of the section, choose any points 2, *b*, 3, in the line KP; from these let fall projectors into the HP, cutting the circle No. 2 in the points 2′2″; 3′3″. Similarly through points 1, 2, *b*, 3, 4, in KP, No. 1, draw projectors parallel to the IL through the axial line *a*′ No. 3, and on them set off from No. 2 the lengths of the lines 2′2″; 3′3″; then a curve drawn through the points thus obtained will give the required projection. As the cylinder is standing vertically on the HP, the plan of No. 3, although having its upper surface inclined to its base, will still be a circle; but as that surface is in section, it has to be cross-lined as shown in No. 4.

Next, let the direction of the cutting plane be KP, No. 5, and an elevation and plan of the section be required, when the part to the left of the cutting plane is removed.

103

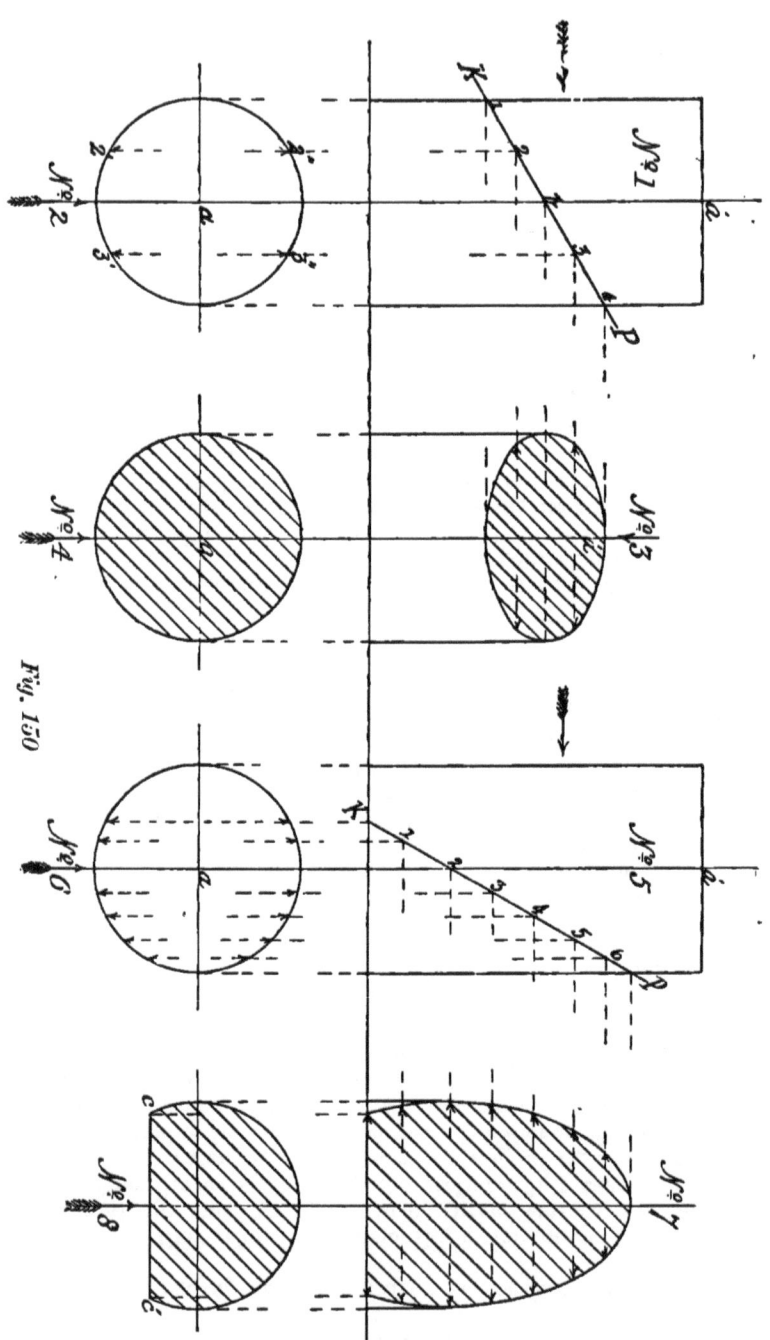

Fig. 150

MECHANICAL AND ENGINEERING DRAWING 105

The portion of the cylinder left in this case is called an *ungula*, and the required elevation and plan of the section are obtained in a similar way to No. 3 and No. 4. First, to find the elevation: choose, as before, a convenient number of points in the line KP; from them let fall projectors, cutting the plan of the cylinder No. 6, as was done in No. 2. Similarly, from the same points draw—parallel to the IL—projectors through the axial line No. 7, and on these set off, in their proper order, the lengths of the corresponding lines found in No. 6; a line drawn through the points thus found will give the elevation or contour of the section made by the cutting plane KP, in its altered position, which is a portion of an elongated ellipse. The plan of this section of the cylinder shown in No. 8, differs little from that of No. 3; the section plane KP, passing out of the cylinder *within* its base at K, cuts off a portion of the base, as shown by the line cc'.

As the projections of all the possible sections of a cylinder are obtained in the same way as the foregoing, further examples need not be given, and we can now proceed to consider those of the cone.

CHAPTER XV

THE PROJECTION OF THE CONIC SECTIONS

47. THE same preliminary remarks may be made in reference to the simple sections of a "cone" as were made in the case of the cylinder. From the definition given of a cone, and its assumed mode of generation, it will be at once seen that any section of it taken through its apex to its base will be a triangle, while a section parallel to its base—or at right angles to its axis—will be a circle. Admitting these as self-evident truths, we have then to define those other sections of the cone not taken in the directions mentioned, and to show how their projections are obtained.

The three most important sections of a cone—other than the two mentioned—are the "ellipse," the "parabola," and the "hyperbola."

If a cone be cut by a plane inclined to its axis, or its base, the section is an *ellipse*, or a regular closed curve having diameters differing in length.

If the section of a cone be made by a plane parallel to its axis, but not coinciding with it, then the section is a *hyperbola*.

If a cone be cut by a plane parallel to its slant side, then the section is a *parabola*. From its definition it will be readily understood that the elevation of a cone when standing with its axis vertical is a *triangle*, and its plan a *circle*. Before attempting the projection of any of its sections, it must first be known how to find the plan, or elevation—either being given—of a point on its surface, as this will materially assist in finding its sections.

In Fig. 151: If No. 1 be the plan of a cone, No. 2 its elevation, and x in No. 1 a given point on its surface, it will be seen from its assumed mode of generation that this point must lie somewhere in the slant side of its triangular generator $a'Ab$, No. 2. Now *every* point in this line—it may be assumed to be made up of points—in its revolution about the axis $a'a$ of the cone describes a circle, and the circle in which x must lie will have a radius equal to the distance between it and a; therefore in No. 1, with a as a centre and $a\ x$ as radius, describe a circle cutting the diametral line AB in x'. Again, through the given points x and a in No. 1, draw the line axc, cutting the base AB of the cone in c. Find in No. 2 the elevation c' of c

106

MECHANICAL AND ENGINEERING DRAWING   107

No. 1, and from it draw the line $c'a'$ No. 2; then this line will be the elevation of ca, No. 1, on the surface of the cone, and it is *in it* that the given point x lies. To ascertain where, get the elevation of the circle drawn through x in No. 1 by a projector from x', cutting the slant side $a'A$ of No. 2 in point x', then the point x, where a line through x' parallel to AB cuts $a'c'$ No. 2, will be the required elevation

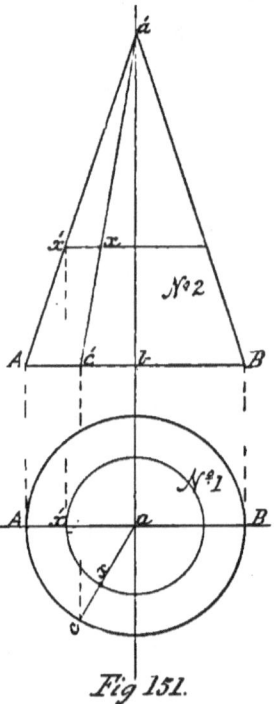

Fig 151.

of the original point x given in No. 1, for the line last drawn may be assumed to be the base of a cone of a lesser height than the given one, x being a point in its base in the same position relatively that c or c' is in the base of the larger or given cone. By a converse process to that here so fully explained, the plan of a point on the surface of a cone may be found from its given elevation. We now proceed to the projection of the conic sections, and as a first problem we take—

Problem 52 (Fig. 152).—*Given the plan and elevation of a cone, to find its sectional projections when cut by a plane at an angle of 30° with its axis.*

Let No. 1 and No. 2 be the given plan and elevation of the cone, and LS, No. 2, the line of section. Draw in the axial line No. 3, and

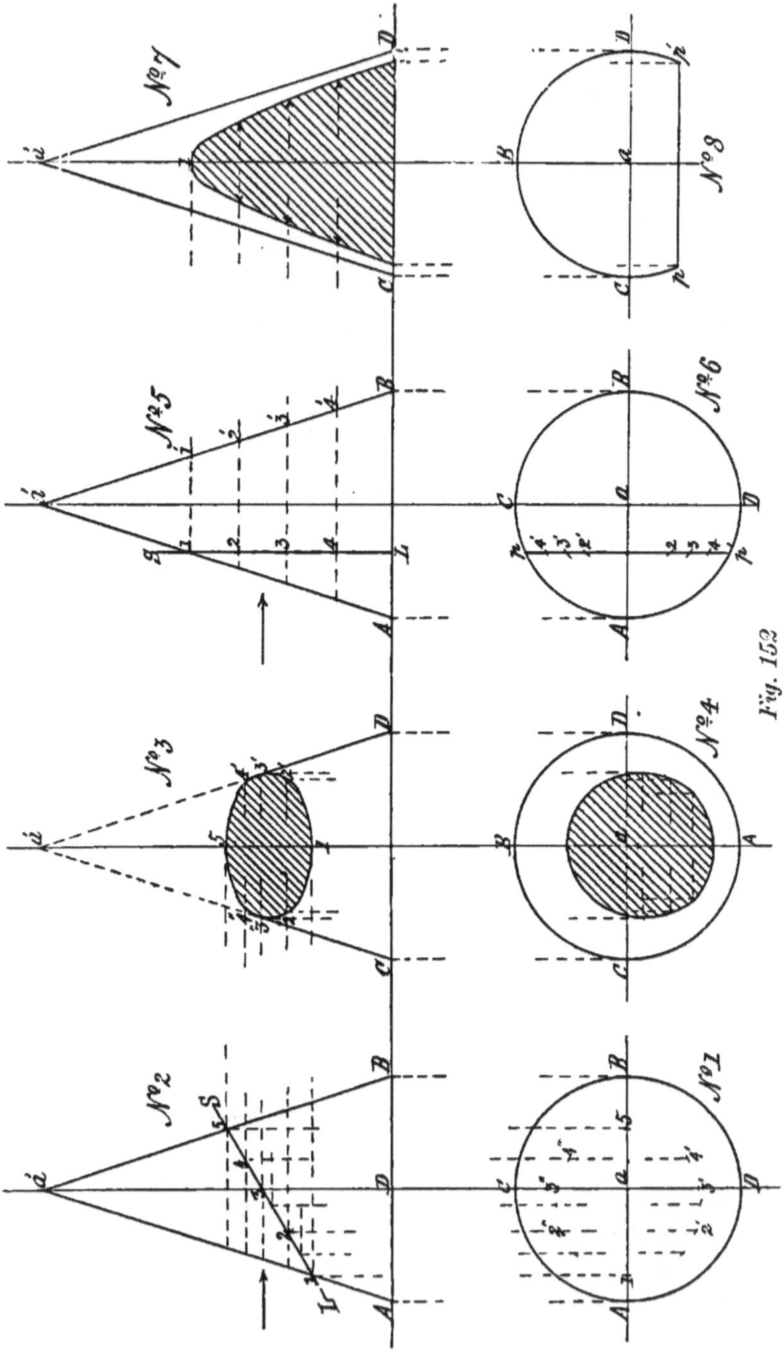

Fig. 152

MECHANICAL AND ENGINEERING DRAWING   109

produce it into the HP. Choose any convenient points, as, 1, 2, 3, 4, 5, in LS, No. 1, and through them, parallel to the IL, draw lines to cut the sides Aa', Ba', of the cone. These, being parallel to its base, will each be the edge-view of a circle or base of a small cone. To find the distance through the cone at the points 2, 3, 4, in LS, No. 2—for this is what is wanted to be known—let fall from them projectors into No. 1, parallel to the axial line aa'; then from a, No. 1, as centre, and with radii equal to half the length of the lines drawn through 2, 3, 4, No. 2, draw arcs, cutting the projectors let fall from these points in 2'2", 3'3", 4'4", No. 1, the length between which is the distance through the cone at their corresponding points in No. 2.

Then, for the contour of the section, project over on to the axial line No. 3, the points 1 and 5 in LS, No. 2, and from the other points 2, 3, 4, draw projectors parallel to the IL through the axis No. 3. On these, set off the lengths 2'2", 3'3", 4'4", taken from No. 1, and through the points thus found draw the curve of the ellipse. With the radius aA, or aB, No. 1, set off from the axial line on the IL, No. 3, the base of the cone in CD; join these with points 3'3", and the required elevation is obtained. The plan of the section obtained in No. 3, will be the ellipse, found by drawing a line through the intersections of the arcs and double ordinates in No. 1. To show this in its correct position as a projection of No. 3, transfer the points and lines in it to No. 4, as shown, and it will then be seen that projectors let fall from the points in the section No. 3 will fall upon their transferred plans in No. 4. A line drawn through these points will be the required plan of the cone when cut by the section plane LS, No. 2. The student must remember that the view No. 3 being at right angles to that of No. 2, the plan No. 4 is necessarily in the same position with respect to No. 1.

Next, let the cone be cut by a plane parallel to its axis, and its sectional projections be required.

48. In this case the projected section will be a true "hyperbola," as the view obtained of it will be directly at right angles to the cutting plane. Let No. 5 and No. 6, Fig. 152, be the given elevation and plan of the cone, and LS, No. 5, the line of section. The simplest way to obtain the required elevation will be to cut up the cone—from the point where the cutting plane enters it—into horizontal sections, and then find where the plane of section enters and leaves their horizontal surfaces. To do this, draw lines at suitable distances apart, as 1 1', 2 2', 3 3', 4 4', parallel to the IL, as shown in No. 5. These will all be edge-views of circular planes perpendicular to the VP. The circle ACBD, No. 6, is the plan of the cone; to find where the section plane passes through it, produce the line LS, No. 5, and it will cut the circle in a line $p\,p'$ parallel to CD. The part of the cone to the left of the section plane may now be supposed to be removed, showing when looked at in the direction of the arrow on the left the exact contour of the section. To find this, all that is required to be known is the width across the *face* of the section at the heights of the circular planes 1, 2, 3, 4, No. 5. Draw in, as in No. 7, the side elevation of the cone. Project over from No. 5 on to its axial line the point 1, where the cutting plane first enters the side of the cone—which will be the highest point in the curve—through points 2, 3, 4, in LS, No. 5; draw lines in No. 7, parallel to the IL or CD. In No. 6, with a as a centre, and the half-lengths of the lines drawn through 2, 3, 4, No. 5, as radii, draw arcs cutting the line pp', No. 6, in the points 2 2′, 3 3′, 4 4′; the distances between these points severally will be the distances across the face of the section at the heights of their corresponding points in No. 5. Set these distances off on the lines drawn in No. 7; then a curve drawn through the points thus found will be the required sectional elevation of the cone, cut by a plane in the position given. This curve is a true *hyperbola*, because the plane of the section of the cone producing it is parallel to the cone's axis and to that of the plane of its projection. The plan of this section, from its position with respect to the HP—viz., perpendicular to it—becomes a straight line, which, combined with the part of the cone not affected by the cutting plane, will appear as shown in No. 8.

Again, let the cone be cut by a plane parallel to its slant side, and the sectional projections, elevation, and plan be required.

Let No. 1 and No. 2 (Fig. 153) be the elevation and plan of the cone, and LS the cutting plane, drawn parallel to its slant side Aa'. Here the *true* form of the section will be a "parabola," but the plane of the section being inclined to that of its projection, or the VP, the view obtained of it, looking in the direction of the arrow, will be its apparent and not its true form; but this point will be explained further on.

To find the sectional elevation, first choose, as before, suitable points, such as 1, 2, 3, 4, in the line LS, No. 1; through them and point S draw lines parallel to AB—the base of the cone—to touch both of its slant sides Aa', Ba'. These lines, as in the last problem, will be edge-views of circular planes. From the points S, 1, 2, 3, 4, L, let fall projectors on to the plan No. 2, parallel to the axial line $a'a$. With the half-lengths of the lines first drawn through S, 1, 2, 3, 4, as radii, and a, No. 2, as centre, cut the vertical projectors let fall from these points in 1′1″, 2′2″, 3′3″, 4′4″. Through the corresponding points in the elevation No. 1 draw horizontal projectors through the axial line $a'a$, No. 3, and on them set off the distances—taken from No. 2—between the points 1′1″, 2′2″, 3′3″, 4′4″; then a line drawn through the points thus found will give the elevation required.

The plan of this section, as shown at No. 4, is obtained in the same way as the plan of the hyperbola No. 6 (Fig. 152)—viz., by transferring to No. 4 the points obtained by the intersection of the arcs and double ordinates in No. 2, and through them drawing the parabolic curve, as shown.

By the same method of procedure as shown in the three foregoing problems, the projection of any possible plane section of the cone can be obtained. Nor is the method confined to the sections made by a plane:

MECHANICAL AND ENGINEERING DRAWING   111

it is equally applicable to those producing a curved sectional surface. As an example in this direction, take the following as a problem—

Problem 53 (Fig. 154).—*Given the front elevation of a cone, and the curve of the line of section, to find the sectional elevation of the same.*

Let No. 1 (Fig. 154) be the given elevation of the cone, and LS the curved line of section. Take, as in the previous problems, any convenient number of points in LS, No. 1, and through them draw lines parallel to the base of the cone. Draw in the plan of the cone No. 2, and let fall into it, parallel to the axis, projectors from the points taken in LS, No. 1. With *a*, No. 2, as centre, draw arcs as before, cutting the vertical projectors from the points in the curve in No. 1 in corre· sponding points in No. 2. Then, to find the elevation of the curved section, draw in the outline of the cone as in No. 3, and the projectors from the points in LS, No. 1; upon these set off the length of the double ordinates previously found in No. 2, and through the points so obtained draw in the closed curve shown in No. 3. The plan of the section found in No. 3 is obtained in precisely the same way as in the previous problem—viz., by transferring the points found in No. 2 to No. 4, and drawing through them the closed curve as shown therein.

49. In a previous paragraph reference is made to the question of the *true* and *apparent* form of the section of a solid. As it is absolutely necessary that the student should distinctly appreciate the difference between these views of a solid, we give in Fig. 155 a graphic illustration, showing wherein the difference exists. The solid chosen for its explanation is the cylinder, it being the simplest for the purpose.

A cylinder—ABCD (Fig. 155)—is placed with the axis perpendicular to, and one of its ends resting on, the HP ; the true form of a section of it is required when cut by a plane SP, inclined to its axis at an angle of 45°.

Now, a *true* section of the solid on the line given should. be such that—supposing the cylinder to be of material substance and cut through on that line—the section of it obtained by projection, if cut out in paper and laid on the cut surface of the solid, should exactly fit it.

In the projected section of a similar solid in Fig. 150, it is evident that if the ellipse obtained were cut out in paper and applied to the solid on the cut surface, it could not possibly fit it, as the majoɪ axis of the projected ellipse would be found to be of a less length than that of the line of section ; to find, therefore, the true section, a view of it must be got directly at right angles to the plane of section, or in the direction of the arrow *y* in Fig. 155.

This view may be obtained in two ways, either by direct projection from the plane of section with the cylinder *standing as it is* in the figure, or by moving the cylinder into such a position as to bring the plane of section SP at right angles to both the VP and HP, and then getting a side elevation of it. The latter of the two ways, as it gives

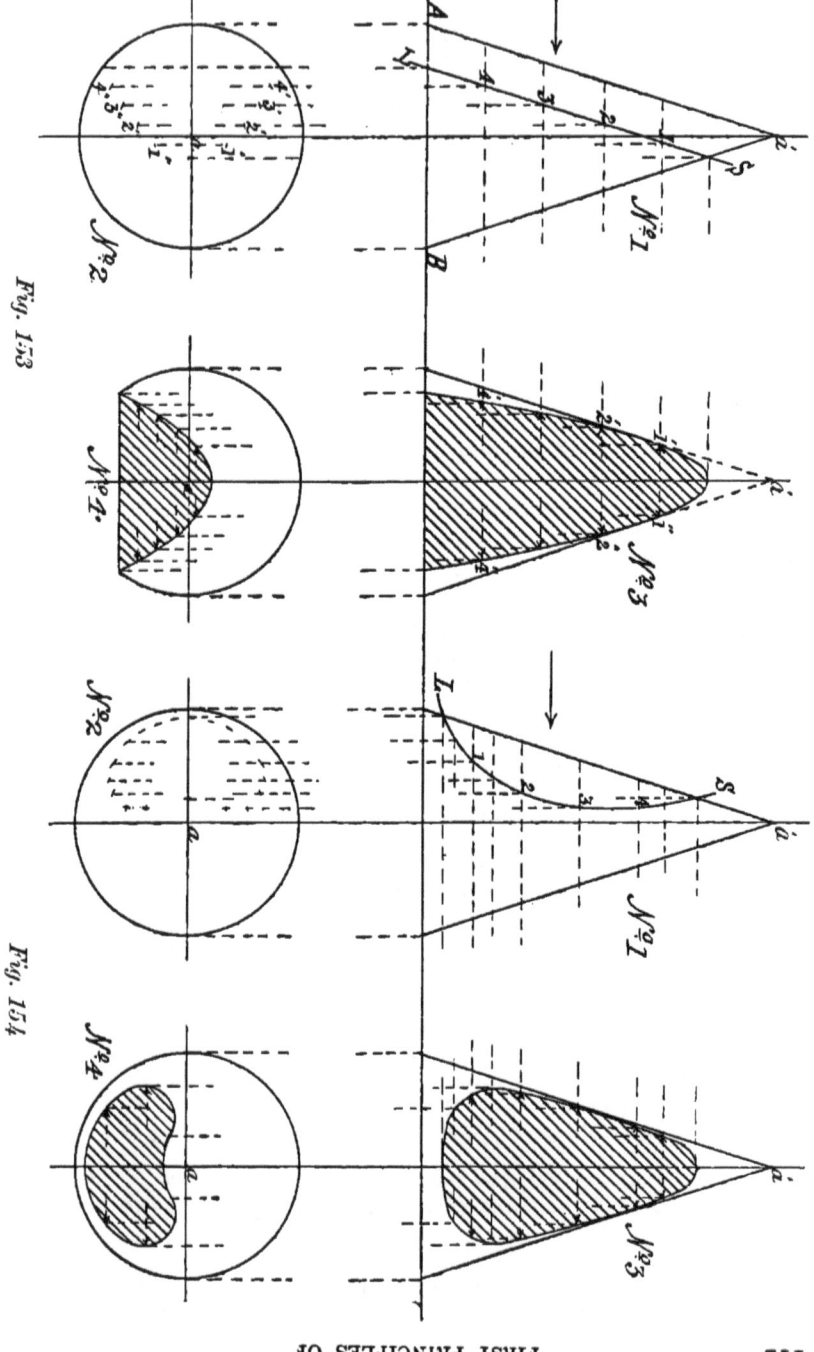

Fig. 153

Fig. 154

MECHANICAL AND ENGINEERING DRAWING   113

the student a fuller insight into the methods of procedure in such a
case, is the one adopted.

Now, as the cylinder is shown as vertical and the section plane
inclined at 45° to its axis, to bring this plane perpendicular to the HP
it must be moved through an angle of 45°. For the purpose of the
projection it is assumed that the *upper* part of the cylinder *above* the
section plane is removed. To bring the lower part into the required
position it must be turned through a vertical arc of 45°. To do this, at
the point D in its base draw indefinitely to the right and left lines
making angles of 45° with the IL. On the one to the right, set off Dp

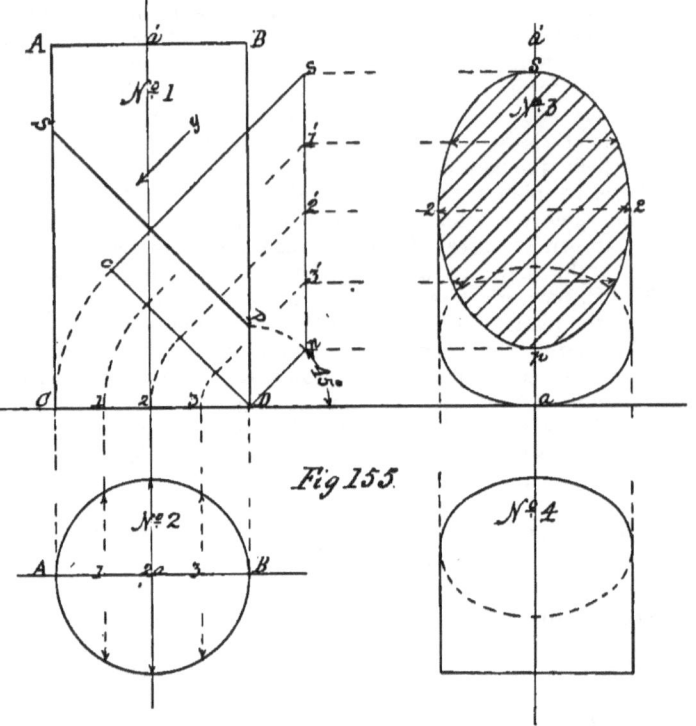

Fig 155.

equal to DP; and on that to the left, Dc equal to DC—the diameter of
the base of the cylinder; at c draw cs equal to CS, the long side of the
cylinder, at right angles to cD; join sp, then Dcsp will be the lower
end of the cylinder when the plane of section SP is vertical, or at right
angles to the VP and HP. The true form of the section can now be
found. First draw in No. 2 the plan of the cylinder in its vertical
position. Divide its diametral line AB into four equal parts in points
1, 2, 3; through these draw double ordinates and projectors to cut CD,
No. 1, in 1, 2, 3. Transfer by arcs points 1, 2, 3 in CD, No. 1, to cD,
and through these last draw lines parallel to cs to cut sp in points

I

114   FIRST PRINCIPLES OF

1', 2', 3'. For the *form* of the section *sp* draw in, in No. 3, the axial line *a'a;* project over to it from *sp*, No. 1, the points *s'* and *p'*, and from 1', 2', 3' in *sp* draw projectors through *a'a*, No. 3. On these set off the length of the corresponding double ordinates drawn through 1, 2, 3, No. 2, and through the points thus found draw the ellipse S2*p*2, No. 3, which is the required *true* form of the section at the line SP, No. 1.

50. As opportunities will occur as we proceed with this part of our subject of obtaining the true form of section of any solid in another way than that just explained, we pass on to the sections of the third of the primary solids with which we are dealing—viz., the "sphere."

Now, as the sphere is generated by the revolution of a semi-circle about its diametral line—or chord—as an axis, it follows that *any* plane section of the generated solid must be a circle, whether it pass through the axis or not, for any and every point in the curved-bounding line of the semi-circle not only describes in its circuit round the axis a true circle, but is at the same time at an equal distance from the centre of the solid which is generated. It follows from this that all the sections of a sphere, by planes passing through its centre, are equal circles, having the same diameter and centre as that of the sphere. Such circles are known as "great circles" of the sphere, and their projections, if taken at right angles to the plane of section, are circles equal to their originals. If, however, a sectional projection of a sphere is required at an *angle* to the plane of section, then the projection becomes an ellipse, and is obtained in the same way as that of any circular plane or disc when at an angle to the plane of its projection. As all such sections of a sphere would in projection be ellipses, only one problem is given in this connection, which will show the difference between the sections of the same sphere when cut by a plane at varying angles to its plane of projection. The problem is—

Problem 54.—*Given the plan of a sphere and the line of section, to find its sectional elevation.*

Let AB, No. 1 (Fig. 156), be the plan of the sphere, and SP the line of section, the axis *a'* or pole of the sphere being perpendicular to the HP. Here SP being parallel to the axis *a'* and to the VP, the section of the sphere by it will be a circle. To find its elevation, first show in that of the sphere AB, No. 2. Project over point 1, in SP, No. 1, to 1 in AB, No. 2; then in No. 2, with *s* as centre and *s*1 as a radius, describe the circle 1, 2, and it will be the sectional elevation required. For SP, No. 1, being parallel to the VP, the circular section made by it will be concentric with the great circle AB, No. 2, and therefore their centres will be coincident; and as SP cuts the great circle AB, No. 1, in points 1 and 2, the elevation of the section made by SP will be a circle, having a diameter equal to the distance between those points, as shown by its projection in No. 2.

Next, let the sphere be turned on its axis in the direction of the arrow in No. 1, until the section plane SP is at an angle—as in No. 3—with the VP, and its elevation be required.

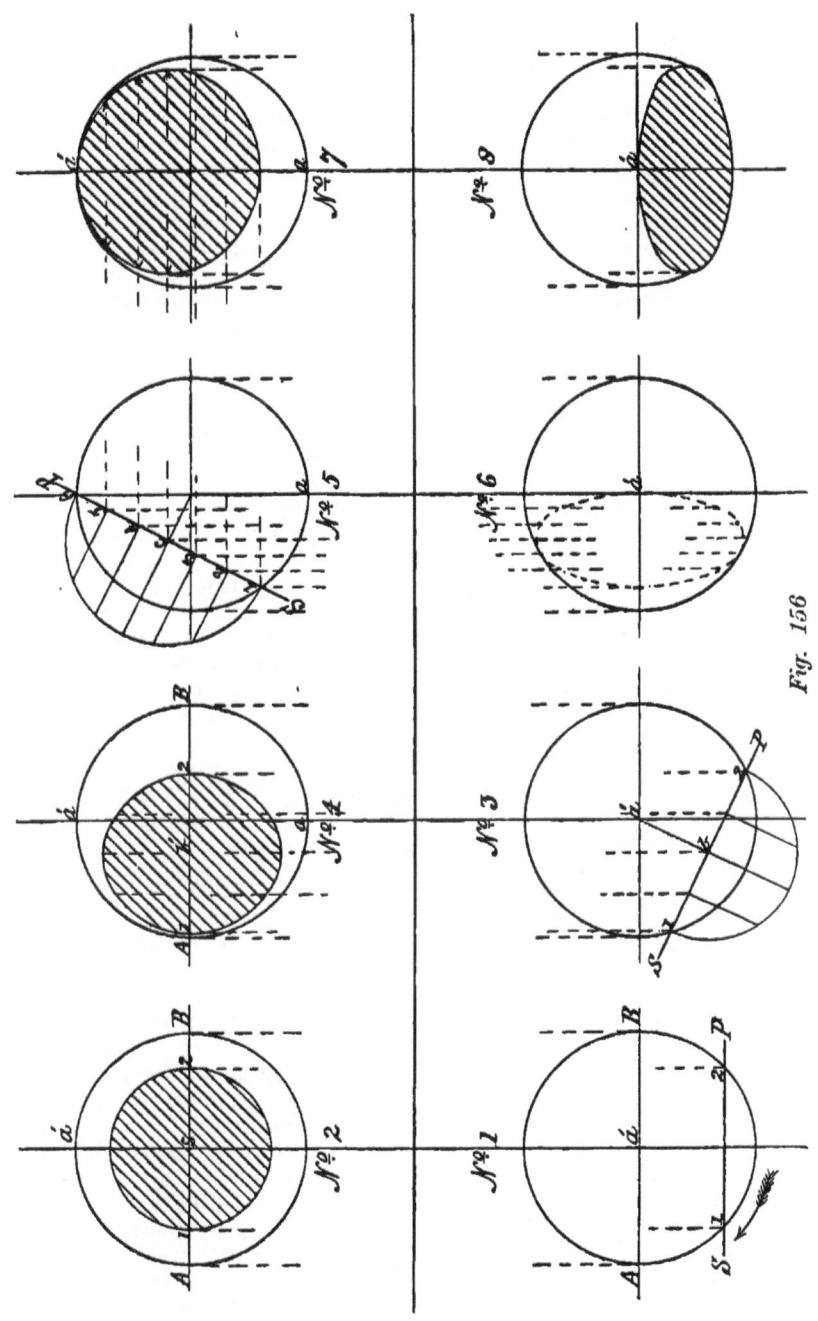

Fig. 156

FIRST PRINCIPLES OF

Here, SP being inclined to the VP, the elevation of the section will be an ellipse. To find it, first get the elevation of the sphere, as in No. 4, having $a'a$ for its axial line. Bisect the distance between 1, 2 in SP, No. 3, in the point b, and through it draw a projector to cut AB, No. 4, in b'; also project over points 1, 2 in SP, No. 3, to points 1, 2 in AB, No. 4. Now the true section of the sphere made by SP is a circle of a diameter equal to the distance between points 1 and 2 in that line, and it is the elevation of this circular section inclined to the VP, as shown, that is here required. It is left to the student without further explanation to find the resulting elliptic section, as was done in Problem 48, Fig. 143.

Again, let the problem be—

Problem 55.—*Given the elevation of a sphere cut by a plane inclined to the HP, to find its sectional projections.*

Let No. 5, Fig. 156, be the elevation of the sphere, and SP the line of section. To find the side elevation, first draw the outline of the sphere, as in No. 7. Then as the true section of the sphere by SP is a circular plane, find by Problem 48 its vertical projection when inclined as in No. 5, which will be an ellipse as shown in No. 7. For the plan of No. 7 in its correct position, proceed as was done in the case of the cone No. 1, Problem 52, by finding the plan of the points in the contour of the section, as in No. 6, and transfer them to No. 8; then a line drawn through the points so found will give, as in No. 8, the required sectional plan of No. 7.

In the same way as here shown can the projection of any possible section of the sphere be obtained, no matter how the line of section be drawn, or whether it is straight or curved, as the same rule holds good in this case as in that of the cylinder and the cone—viz., Find the actual distance *across* the face of the cut surface of the section, from any points on the surface of the solid—in the line of section—*nearest* the eye, to those directly behind them on the opposite side *farthest* from the eye, and the problem of its projection is solved, for the points denoting the distances thus found have only to be put in on the paper in their proper places, and a line drawn through them will give the sectional projection sought.

51. Having shown how the sectional projections of the three primary solids—the cylinder, the cone, and the sphere—are found, we now pass on to the consideration of those subsidiary solids of revolution which are generated by *parts* of the sections of the primary ones.

The particular solids of this class which assist in giving shape to machinery details are the "spheroid," the "conoid," the "cylindroid," the "spindle," and the "oval"; and they are thus defined :—

A *spheroid* is a solid generated by the revolution of a semi-ellipse about one of its diameters. If about its *longer* one, it is a *prolate* spheroid; if about its *shorter* one, an *oblate* spheroid.

A *conoid* is a solid generated by the revolution of half of a conic section about its axis, and becomes an "ellipsoid," "paraboloid," or

MECHANICAL AND ENGINEERING DRAWING   117

"hyperboloid," according as its generating figure is half an ellipse, parabola, or hyperbola.

A *cylindroid* is a cylindrical solid having elliptical ends.

A *spindle* is a solid generated by the revolution of an arc or portion of a curve cut off by a chord or double ordinate, around which it revolves as an axis.   It is circular, elliptic, parabolic, or hyperbolic, according as the generating arc, or curve, is a portion of a circle, ellipse, parabola, or hyperbola.

An *oval* is an egg-shaped solid, and may be conceived to be compounded of half a sphere and a like portion of an elliptic conoid.

As all the above-defined solids are assumed to be generated by the revolution of a particular figure about an axis, it follows that a *cross-section*—at right angles to its axis—of any one of them will be a circle, while a section *through* its axis will be a similar figure to that of which its generator formed a half-part.   An oblique section *through* the axis of all of them would be a figure of elliptic form, while one which did not cut the axis would partake of the form of its axial section.   Under these circumstances, as the finding of any required section of one or all of them merely involves the application of methods of procedure already explained, it is unnecessary here to further elaborate them.   All that is required by the student for the mastery of this part of our subject, is to thoroughly appreciate the exact form and mode of generation of the different solids enumerated.   This attained, there will be no difficulty in the solution of any problem that may arise in connection with their projection.

Before concluding this chapter, and with it the subject of the " Projection of Solids with Curved Surfaces," it will be necessary to show, how the principles explained in Chapter XI., on the " Lining-in of Drawings in Ink," are to be applied to representations of this class of solids.

It has been shown in Chapter XI., that the projections of rays of light, in the direction they are assumed (in Mechanical drawing) to fall upon the object illumined, are parallel right lines in plan and elevation, making an equal angle of 45° with the intersecting line of the planes of projection.   To determine then how to line-in the representation of a curved-surfaced solid, we have to find how the light falls upon it.

Now to do this, let the circle *cedb* in No. 1, Fig. 156a, be the plan of a cylinder, and the rectangle ABCD No. 2 its elevation.   Through *a*, the axis in No. 1, draw lines at an angle of 45° with the IL, cutting the circle in points *b* and *c;* and through *b* draw a line parallel to *ca*. Also through *b* and *c*, in No. 1, draw lines in No. 2, perpendicular to the IL, and produce them to meet AB, the top-end of the cylinder in *c* and *b*.   Now the lines drawn through *b*, and *ca* in No. 1, are the plans of two planes of rays of light ; the former touching the cylinder throughout its length in the line *bb'* ; and the latter striking it in the line *cc'* (No. 2).   The line *bb'* is known as the "line of shade," as it divides the cylindrical surface *in the light* from that part which is in *the shade*, and is moreover the darkest part of the visible surface.   The brightest part of the cylindrical surface will be the "line of light" *cc'*, as it is here that the light shines directly upon it.

FIRST PRINCIPLES OF

It follows from the foregoing considerations, that in lining-in the representation of a cylindrical surface in ink, *fine* lines only should be used, as either a medium or shadow line would be inappropriate, the lightest, and darkest, parts of the surface of such a solid always falling *within* the outlines of its representation, as shown at *bb'*, and *cc'*, in the elevation No. 2, Fig. 156*a*. The same rule also applies to either conical or spherical surfaces, as shown in Nos. 4 and 6, in the figure where it is seen that the "lines of shade"—obtained in the same way as in the case of the cylinder—are well within the outlines of the solids, and must therefore be put in, in fine lines.

It would, however, be correct to shade-line the *end* of a cylinder, or the base of a cone—if inverted—as a part of their edges would throw a shadow. To determine how much would do so: in No. 1, Fig. 156*a*, produce the line drawn through *b*, *a* to *e ;* and that through *c*, *a* to *d*. Now as all the rays of light falling on the cylinder, and its upper end AB, are parallel to the one drawn through *c*, *a*, *d*, it follows, that the semi-circular edge *ecb* in No. 1 will be directly in the light, and will cast no shadow, while the opposite similar edge *edb* will do so, either on the HP or VP,—according to its position with respect to those planes—and must therefore be shade-lined. But as the semi-circular edge *edb*—of the cylinder end—would cast different intensities of shadow, the shade-line from *e* through *d* to *b* must be of varying thickness (as shown in the diagram), its thickest part being at *d*.

If the cylinder be hollow—as represented in No. 1—then that part of its inside top-edge, or the semi-circle 1, 2, 3, will throw a shadow on the opposite inside surface, and must be shade-lined as shown, the thickest part of the line being directly opposite the similar part of the outer semi-circle *edb*. In the sphere and other entirely curved-surfaced solids, no shade-lining is admissible in inking them in, fine lines being the only proper ones to be used for the purpose.

MECHANICAL AND ENGINEERING DRAWING   119

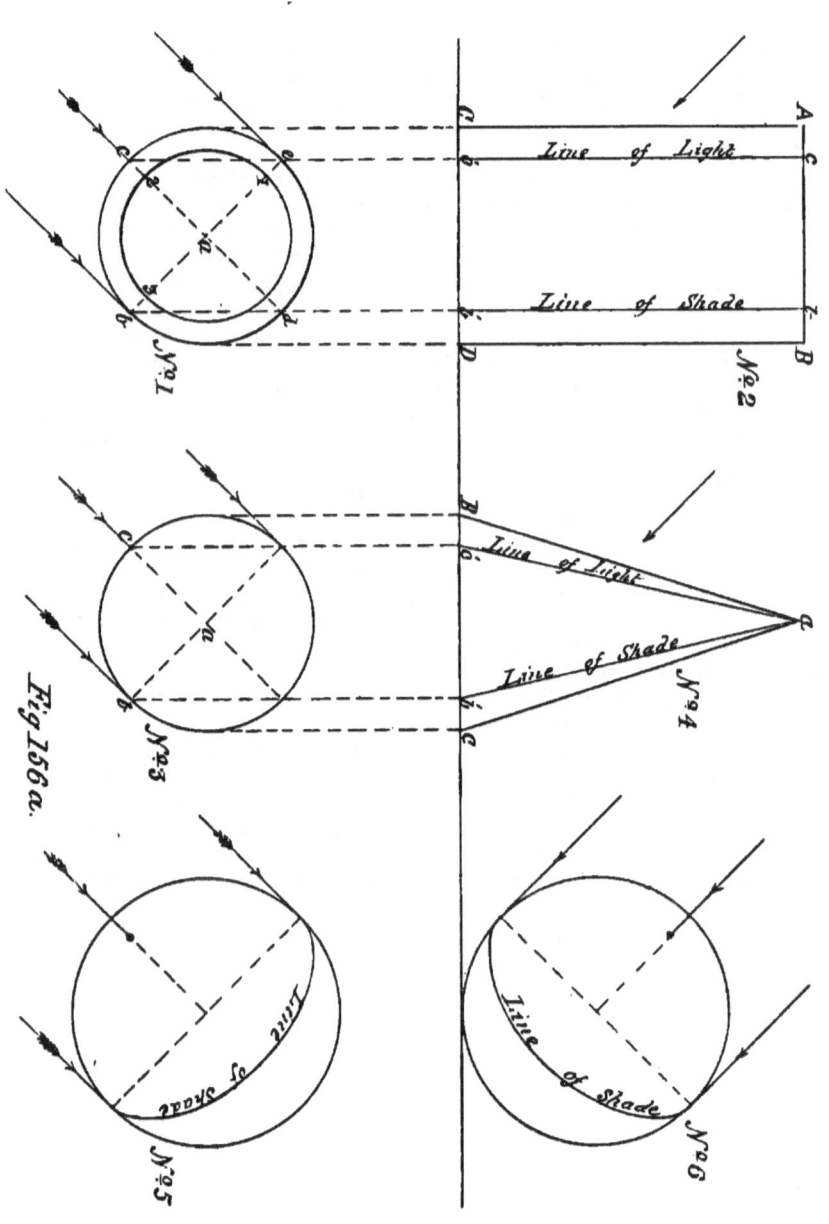

Fig. 156a.

CHAPTER XVI

THE PROJECTION OF OBJECTS INCLINED TO THE PLANES OF PROJECTION

52. FOR the more immediate purpose of representing anything in process of manufacture in the workshop, it will be readily understood, from what has already been explained in these chapters, that the only drawings absolutely necessary to the *workman* are those in which the plans, elevations, and sections of the objects he is called upon to make are correctly shown. But as many of the elements which go to make up a machine, or engine, are often inclined to the vertical and horizontal positions, it is essential that the *draughtsman* should know how to draw the plan or elevation of an object, however intricate, when *inclined* to the VP or HP, at any angle other than a right angle. Before showing how the principles of projection are applied in this part of our subject, we would guard the student against accepting as true, the notion entertained by some, that it is a different kind of drawing to that which has preceded it, and that a knowledge of it is more difficult to acquire. The projections of an object inclined to the planes of its projection, certainly differ in *appearance* to those of it when not so placed, but the method of finding such projections involves no new departure in principle, as the "planes of projection"—the VP and HP—are still in the same relative position as before—viz., at right angles to each other—although the "object" is *inclined* to both. The "projections," therefore, of an object so placed, are nothing more than its appearances when viewed at right angles to those planes.

To show this graphically, let A, No. 1 (Fig. 157), be the plan of a wooden block, with its *under* side resting on a flat plate *abcd*, lying on the HP, and its sides inclined to the VP, or IL, as shown. A front elevation—or view in the direction of the arrow *e*—of the block in this position would be B, No. 3. Now, assume the plate—with the block upon it—to be raised through an angle, moving on its edge *ab*, as a hinge; the elevation of the block in this position would then appear like C, No. 3; for its *upper* face would become more and more visible as the angle made between the plate on which it rests and the HP increased, until the plate had attained a vertical position, when it would again appear as shown in plan at A, No. 1; but as *an elevation*. We therefore see that its "appearance" when inclined to the planes

120

MECHANICAL AND ENGINEERING DRAWING   121

of its projection, although radically different to either its plan A, No. 1, or its elevation B, No. 3, is still nothing more than an ordinary projection of the original object when in its changed position. This is further shown in No. 2 (Fig. 157), where a side view is given of the plate with the block upon it, such as would be seen when looked at in the direction of the arrow *f*, No. 1. Then by the aid of a few projectors from No. 2 and No. 1, it is seen that C, No. 3, and D, No. 4, are just ordinary projections, differing little from those which have preceded them, and certainly no more difficult to obtain than some which the student has already projected or drawn. As the projections of the object given in this figure form a problem for solution later on, we defer showing how they are exactly obtained until we have, as in previous instances, explained how the projections of simple "figures" in similar positions are found.

53. Now, to facilitate the process of finding the projections of an object when inclined to the VP and HP, we have to introduce into the construction an imaginary inclined plane interposed between the VP and HP, on which the object whose projections are required is assumed to rest. This plane is represented in Fig. 157 by the line OP, No. 2, and is the only new feature in our subject of study, it being not a fixed plane like the VP or HP, but one that may assume any desired angle with those planes. Such a view of an object as is shown in No. 2 (Fig. 157) could of course be given without the assistance of the line OP; but as it determines the inclination of the object to the vertical or horizontal, it is of material help in finding its projections, as will be seen as we proceed.

Throughout the problems that will be given in this part of our subject the *original* object whose projections are required is to be always understood as resting on a horizontal plane, and the view first given of it, is its *plan* or horizontal projection, such as A, No. 1 (Fig. 157), which shows all that would be seen of such an object when looked at from above.

Now, to find the projection of a point situate in a plane which is inclined to the VP and HP, we must know the *angle* made by the plane—in which the point lies—with either the VP or HP. Our first problem is—

Problem 56.—*Given the plan of a point, it is required to find its elevation when the plane in which it lies is inclined to the HP at an angle of 30°.*

Let *a* (Fig. 158) be the plan of the given point, IL being the intersecting line. At O draw a line OP, making with the IL an angle of 30°. Through the point *a* draw a projector into the VP. Set off on OP from O, a distance O*a'*, equal to that which point *a* is from the IL. Through *a'* draw a line parallel to the IL, cutting the projector from *a* in plan in point *a"*; then *a"* is the required elevation. In the same way, the elevation of a *line* lying on a plane inclined to the VP and HP is found, for we have only to find the vertical projection of its ends, and join them by a right line, and the desired elevation is obtained.

Problem 57.—*Given the plan of a line, to find its elevation when the plane in which it lies is equally inclined to the VP and HP.*

Let AB (Fig. 159) be the plan, IL being the intersecting line as before. As the plane on which the line AB is to lie equally inclines to the vertical and horizontal, its angle with either will be one of 45°. Therefore, at any point in the IL, say O, draw a line OP at that angle with it. Now OP, we have before said, is a *side* view of the inclined plane on which the object—in this case a line—is to lie; the view we shall obtain of the object in elevation will be one looking in the direction of the arrow *x*, and the position of the VP with respect to OP can be either in front of or behind it. If in front, it would be represented by a line drawn through O perpendicular to the IL; if behind, then a similar line through P. Generally, OP is assumed to be interposed *between* the VP and HP, and therefore a side view of the three planes would be correctly represented by the right-angled triangle PxO in the diagram; the vertical projection or elevation of the inclined object being made on Px or the VP, and the horizontal one, or plan, on Ox, or the HP.

We can now proceed with the solution of the problem before us. Having the inclination of the plane on which the line AB is lying, to find a front elevation of it we require to know the exact position of its two ends with respect to the *foot* of the plane on which it is lying. Now, the intersection of two planes is a straight line; therefore, as the inclined plane OP in the figure intersects the HP in O, the plan of that intersection will be a straight line parallel to the VP or the IL; and assuming the VP to be as far from the foot of the inclined plane OP as *x* is from O, set off from the IL a distance equal to Ox, and through the point so found draw the line DL parallel to the IL, and it will be the plan of the intersection of the HP with the inclined plane OP. It is from this line that all measurements of points in the *plans* of original objects have to be taken, when giving their elevations on the inclined plane OP. Then to find the required elevation of the line AB, set off from O on the line OP in *b* and *a*, the distances *a*A and B*b*, that its ends A and B are from DL, and through them parallel to the IL draw projectors to cut those drawn from A and B in the plan in the points A′B′; join these by a right line and it will be the elevation required.

Then, as lines are the boundaries of "figures," the projection of a figure in a similar position with respect to the VP and HP is obtained in the same way. One problem in this connection will be sufficient to show the application of the principle. Let it be this—

Problem 58.—*Given the plan of a triangle, and the inclination of the plane OP ; to find the elevation of the triangle.*

Let ABC (Fig. 160) be the plan of the triangle, and the inclination of the plane OP be 30° with the HP. Draw in the line OP at the given angle with the IL and the line DL as in the last problem. Find the position of the angular points ABC on the plane OP, and through

MECHANICAL AND ENGINEERING DRAWING   123

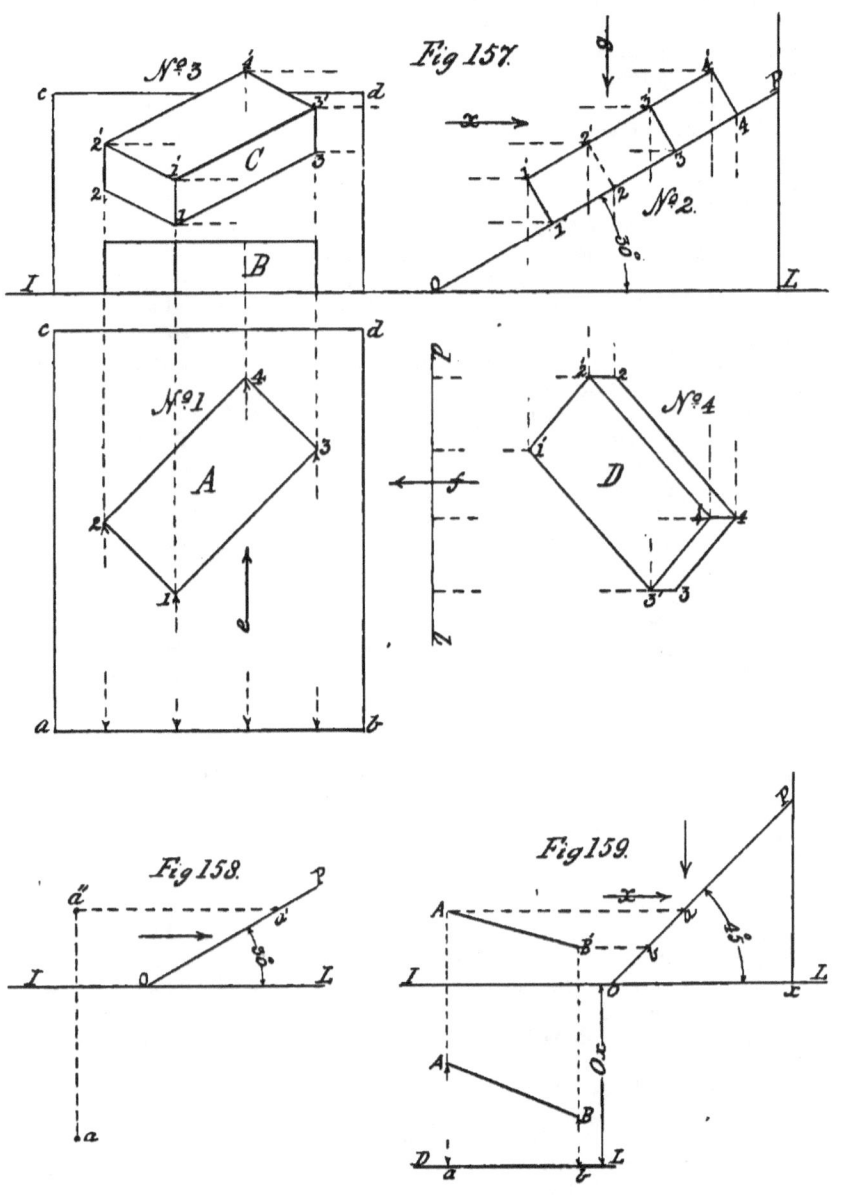

them draw projectors parallel to the IL, cutting those drawn through the corresponding points in the plan. The intersections of these projectors A'B'C' will be the elevations of the angular points of the triangle; join these by right lines and the triangle ABC' will be the elevation required.

If the *plan* of the original figure in its new position be required as well as its elevation, this is found in a similar way. Let the problem be—

Problem 59.—*Given the plan of a rectangular plate and the inclination of the plane OP; to find its projections.*

Let A, No. 1 (Fig. 161), be the given plan of the plate, and the inclination of OP be 30° with the HP. Draw OP, No. 2, at the given angle with the IL, and the line DL as before. Number the corners of the plate 1, 2, 3, 4, as shown in No. 1. Find the position of these points on OP, No. 2, and from them obtain, by projection, the elevation of the plate shown at B, No. 3. To find the plan of the plate in its inclined position, let fall projectors from points 1, 2, 3, 4, in the line OP, No. 2, perpendicular to the IL; then the *corners* of the plate in plan will lie in these projectors, and the problem is to decide where. The line DL, below the original plan of the plate at A, No. 1, is, we have shown, the plan of the intersection of the inclined plane OP with the HP; then a line *dl*, drawn through O, No. 2, at right angles to the IL, as shown in No. 4, will also be a plan of the same intersection. As the distances of the points 1, 2, 3, 4, in A, No. 1, from the line DL, were set off on OP, No. 2, the position of this plane—with the plate upon it—will be at right angles to that of A, No. 1; therefore, its plan No. 4—or the view of it looking in the direction of the arrow *x*, No. 2 —will be in the same position with reference to A, No. 1. Then if the points where the projectors drawn through 1, 2, 3, 4, in A, No. 1, cut the IL, be transferred to the line *dl*, No. 4, and lines be drawn through them at right angles to that line, their intersections with the projectors let fall from 1, 2, 3, 4, No. 2, will give 1, 2, 3, 4, No. 4; join these by right lines, and the required plan of the plate in its inclined position is obtained.

54. As the projection of any plane figure, inclined to both the planes of projection, is found in the same way as here shown, we can without further example proceed with the application of the principles involved, to the projection of *solid* objects similarly placed. Now, a solid object, such as the block of wood taken for explanatory purposes in Fig. 157, has not only length and breadth, but thickness also, and *each* of these dimensions has its representative lines in the projections that are obtained of it, when it is inclined to both the VP and HP. The actual thickness of an object—or its measurement from surface to surface—is the distance through it at right angles to the surface from which it is measured. If the thickness be represented by a line, then the line will be perpendicular to that surface, and its projections will depend upon its position with respect to the planes of projection.

In Fig. 162, if the plan of a line be represented by the point *a*, then, when the plane to which the line is assumed to be perpendicular

MECHANICAL AND ENGINEERING DRAWING

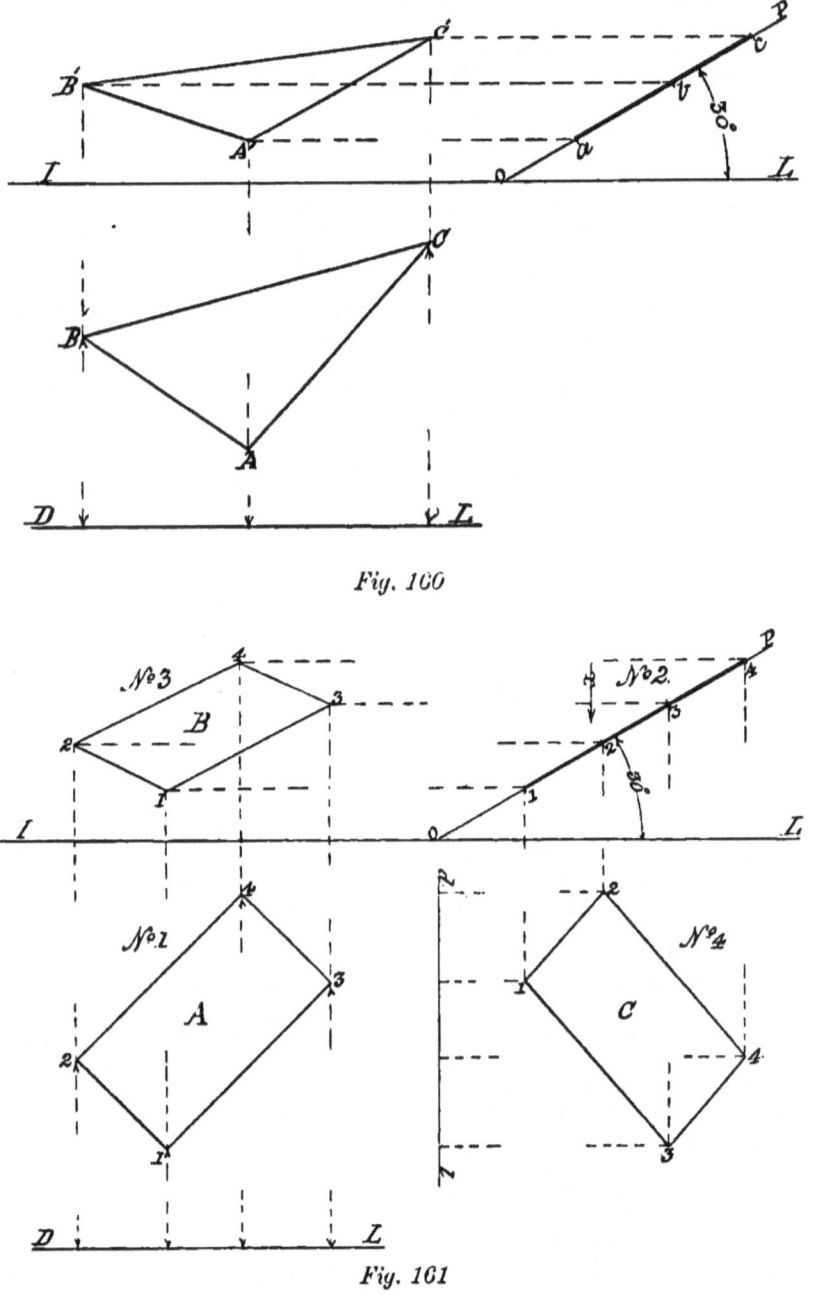

Fig. 160

Fig. 161

126   FIRST PRINCIPLES OF

is inclined, say, 30° to the horizontal, its side elevation on that plane
OP will be $a'b$.   To find its front elevation, we draw a projector through
a perpendicular to the IL, and from a' and b—the two ends of the line
—projectors parallel to the IL, to cut the one through a in $a''b'$; then a
line joining these points is the front elevation of the original line $a'b$
when inclined to both planes of projection.   To apply this reasoning to
the projection of such an object as the rectangular block of wood in
Fig. 157, proceed as follows:—First draw in its plan as at A, No. 1,
and number the four corners as shown.   Each of these will be the end
of a line perpendicular to the HP, of a length equal to the thickness of
the block.   On OP, No. 2, drawn at an angle of 30° with the IL, give
an elevation of the block, looking at it in the direction of the arrow f,
remembering the position of its corners and top surface with respect to
the plane on which it is assumed to be standing.   Having drawn in
this view, proceed to find by projection that of C, No. 3.   This is best
and most correctly done by numbering both top and bottom ends of
the lines representing the angular corners of the block in No. 2.   The

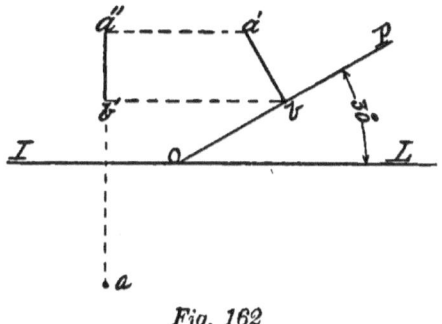

Fig. 162

intersection of horizontal projectors drawn through these points, and
their corresponding vertical ones in No. 1, will give the position of all
the corners of the block that will be seen looking in the direction of
the arrow x in No. 2, and the joining of these by straight lines, as
shown in C, No. 3, will give the required elevation.

To find the "plan" of the block when inclined as shown, let fall
vertical projectors from all the corners of it seen from *above* in No. 2,
when looking in the direction of the arrow y; transfer the points
where the vertical projectors from 1, 2, 3, 4, No. 1, cut the IL, to the
projector drawn through the foot of the inclined plane at O, as shown
in Fig. 161 previously, and through these draw projectors to cut those
let fall from 1, 2, 3, 4, No. 2; then the intersections of these projectors
will be the plans of the corners of the block seen from above it, which,
on being joined by straight lines, will give D, No. 4, the plan of the
block, when inclined in the position shown in side elevation at No. 2,
and front elevation at No. 3.

55. As a test of the correctness of the two last-found projections,
all the lines in the original object A that are parallel to each other

MECHANICAL AND ENGINEERING DRAWING   127

should be parallel in their projections, for if they are not, there is some inaccuracy in projecting over corresponding points in the different views, as no alteration of *position* of the original object, with respect to the planes of its projection, can in any way affect its *form*, or the relative position of its *surfaces*, from whatever direction it may be viewed. These deductions apply, of course, to all "original" objects, but they require to be more particularly remembered here, as in no other part of the subject is there the same likelihood of making a false projection of an object, as when it is inclined to *both* the planes of its projection.

As no better examples of plane-surfaced solids, as objects for projection in this part of our subject, can be chosen than those whose forms the student is already familiar with—viz., the cube, prism, pyramid, etc.—they will be taken in the same order as before, and the first problem is—

Problem 60 (Fig. 163).—*Given the plan of a solid cube, with one of its faces upon the HP ; to find its plan and elevations, together with a sectional plan and elevation, when the plane on which it rests is inclined to the horizontal at an angle of 30°.*

Let the square 1, 2, 3, 4, No. 1, be the given plan of the cube, IL being the intersecting line of the VP and HP, and DL a datum line parallel to the IL, at any convenient distance below the plan of the cube. At O, say in the IL, draw a line OP, making with the IL an angle of 30°. Set off on OP, from O, in points 1', 4', 3', the distances that points 1, 4, 3 in No. 1 are from the line DL, measured at right angles to it. Through 1', 4', 3', No. 2, draw lines perpendicular to OP, and on that through 1' set off from OP a height 1'1, equal to a length of the side of the cube. Through 1, draw a line parallel to OP, cutting the perpendiculars to it at points 4 and 3 ; and the view No. 2 will be a side elevation of the cube—looked at in the direction of the arrow *x* in the plan—when inclined to the HP at 30°.

To find a front elevation of the cube at the same inclination, we have to consider what parts of it will be seen in such a view. This will evidently include the two inclined front faces and a top face, or what would be seen looking in the direction of the arrow to the left of No. 2. To find the projection of these faces, project over points 1, 1' ; 4, 4' in No. 2, on the projectors drawn through points 1, 2, 4, No. 1, cutting them in points 2, 2' ; 1, 1' ; 4, 4', No. 3 ; join these by right lines as shown, and the projection of the two front faces of the cube will have been found. To find the top face, project over in like manner point 3 in No. 2, to cut the vertical projector through the corresponding point in No. 1, and it will give point 3 in No. 3 ; join this by right lines with points 2 and 4, and the required front elevation of the cube in its inclined position is obtained.

To find the plan of No. 2, or a view in the direction of the arrow shown above it, first find the plan of the top face of the cube, by letting fall vertical projectors from points 1, 4, 3, in it, into the lower plane, or HP ; intersect these by horizontal projectors from points 1, 2, 3,

128   FIRST PRINCIPLES OF

No. 1, in points 1, 2, 3, 4, No. 4; join these by right lines, and the
plan of the top face of the cube is found.   For the plan of its two
inclined faces, seen from above, produce the horizontal projectors drawn
through the three corners 2, 3, 4 of the top face—just found—and by
vertical projectors from points 4' and 3', in No. 2, cut the produced
horizontal ones in points 2', 3', 4', No. 4; join points 2, 2'; 3, 3'; 4, 4';
and 2', 3'; 3', 4', by right lines as shown, and No. 4 is the required plan
of the cube in its given inclined position.

56. Before proceeding to show how the *sectional* projections of the
cube are found, it is advisable here—to save repetition later on—to
explain the reasoning which has directed the process of solving the
first part of this problem.   Taking the "original" object—the cube—
as the first subject of thought, we have to note its special form, and the
shape and relative position of its six faces and their bounding edges, as
upon the correct realization of these features depends in a great measure the truth or falsity of its projections.   For whatever relative position its sides have to each other in the original, they cannot, as long as
they remain uncut, have any other than the same relationship in their
projections.   A second consideration is the position of the object, and
its faces, with respect to the plane on which it is assumed to be standing; and a further one is that of the inclination this plane eventually
assumes with respect to the planes of projection.

Now, the cube is standing with one of its faces on the HP, thereby
making its four corner edges, figured 1, 2, 3, 4, No. 1, in the diagram,
at right angles to that plane, and the top face of the cube parallel to
it.   It follows from this that however much the plane on which the
cube rests is inclined to the horizontal, the position of the faces and
corner edges of the cube with respect to that plane will remain exactly
as before its inclination.   Therefore, in setting the cube on the inclined
plane OP in the diagram, all its vertical edges will be perpendicular to
OP, and its horizontal ones parallel to it.   Then as to its projections
No. 3 and No. 4, although the cube itself in its new position No. 2
is inclined to the VP and HP, this has not in any way altered the
relative position of the bounding edges of its faces in the projections,
as the lines 1, 1'; 2, 2'; 3, 3'; 4, 4', are still parallel to each other, as
well as those representing the top and bottom edges of the cube.

To assist the student to a more complete realization of the position
of the "planes of projection" with respect to the original object, in
Fig. 163 let a line be drawn perpendicular to the IL through point 3'
of the cube in No. 2.   This line would then represent an end elevation
of the VP, with the corner 3' of the cube touching it.   If, then, the
inclined plane OP, with the cube upon it, together with this vertical
plane, be imagined to swing round on the HP through 90°, the view of
the cube shown at No. 3 would be identical with its appearance on the
inclined plane when viewed at right angles to the VP.

57. With the foregoing explanation of the reasoning applied in
finding the projection of the cube in its uncut state when inclined to
the VP and HP at a given angle, there should now be no difficulty in
obtaining its sectional projections required by the problem, as these
are nothing more than similar views to those already found, with the

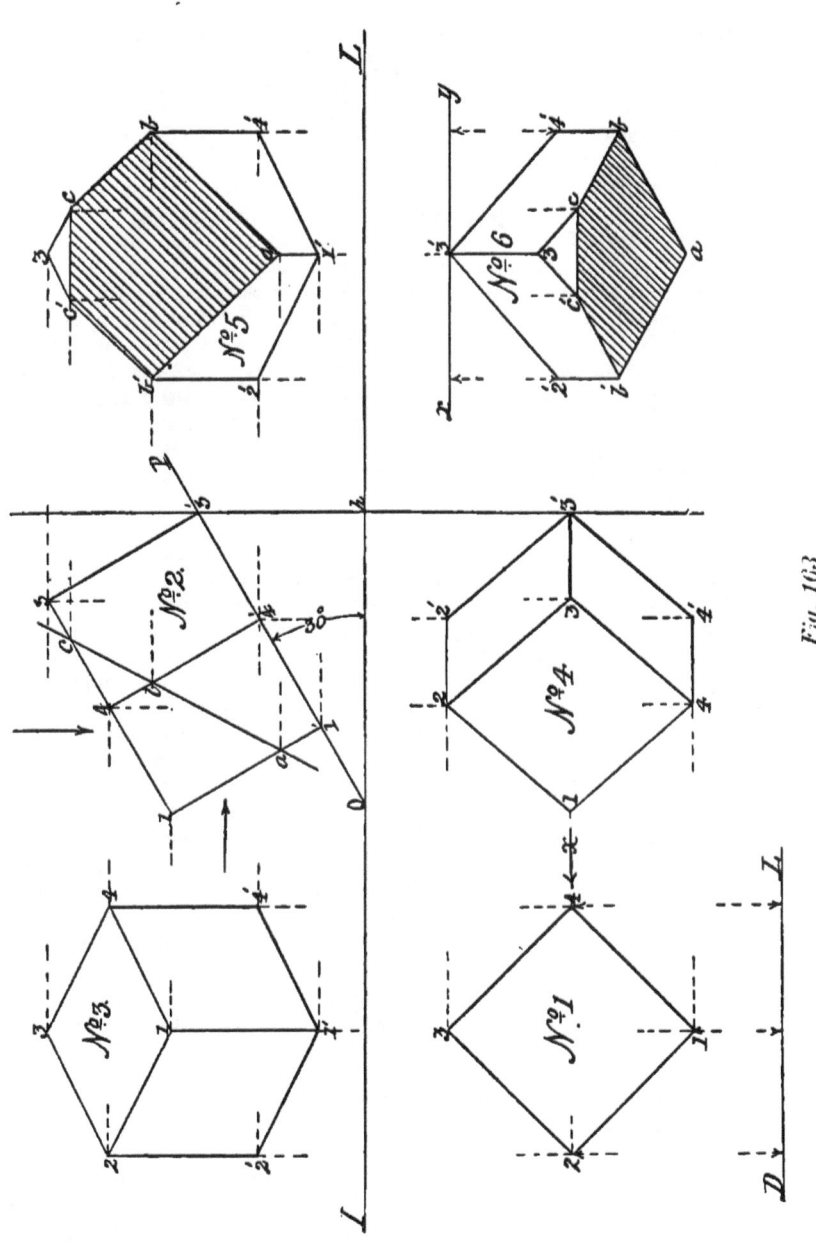

Fig. 163

130   FIRST PRINCIPLES OF

exception that a portion of the solid is assumed to be removed, exposing
the cut surface.

The fact of the cube being in the same *position* as before, reduces
the problem to the finding of the projections of its exposed cut surface,
when the part to the left of the cutting plane is removed.   For the
sectional elevation, draw in first—in faint lines—to the right of No. 2 a
similar view of the cube to that given in No. 3.   The cutting plane *ac*,
it will be seen, passes through *five* of the faces of the cube, entering it
at *c* on its top face, passing through its two back and front faces, and
leaving it at *a* in its foremost edge.

Now, the position of the cube in No. 3, with respect to the VP, is
that of having its two front and back faces making equal angles with
it, and it follows from this that a plane passing through the cube, as
ac is assumed to do, will cut *opposite* faces of it in parallel lines.   Therefore to find the projection of the section in No. 5, made by *ac* in No. 2,
through *c* draw a horizontal projector to cut the back top edges of the
cube in *c′ c*, No. 5, and similar projectors through *b* and *a*, the former to
cut the side edges 2, 2′ ; 4, 4′ in *b b′*, and the latter the front edge in *a*.
Then join by right lines in No. 5 the points *ab*, *bc*, *cc′*, *cb′*, *b′a* thus
found, and the space enclosed by them will be the sectional surface of
the cube cut by the plane *ac*.   Put in, in full lines, the remaining
uncut part of the cube, and the complete view, as at No. 5, thus
obtained will be the required sectional elevation of the cube.

Then as this last-found view is one looking at the cube in No. 2, in
the direction of the arrow on its left, when the portion on that side
of the cutting plane *ac* is removed, the plan now required is a view
of this remaining part when looked at from above, and as the position
of No. 2 is at right angles to that of No. 5, the plan of this latter,·
which is what has now to be found, will be in the same position with
respect to No. 4.

Now, the simplest way to obtain this view is to make use of the
assumed VP, previously drawn in No. 2, touching the corner 3′ of the
cube.   As No. 5 is a view of the cube directly at right angles to that
at No. 2, this assumed VP will be *behind* it in No. 5, and a plan of
it in that position will be a line in the HP parallel to the IL.   Draw
such a line as *x y*, No. 6 ; then all the points and lines seen in No. 2
from *above* can be set off from this line.   The back lower corner of the
cube is touching it, and we know its position is behind 1′, No. 5 ;
therefore let fall first a vertical projector from this point to *x y*, No. 6,
in point 3′, and similar projectors from points 2′ and 4′.   Set off from
x y on the projector through 3′, and similar projectors from points 2′
and 4′.   Set off from *x y* on the projector through 3′, the distance
3′ 3 equal to that of point 3, No. 2, from the assumed VP, and on
those let fall from 2′ 4′, No. 5, in points 2′ *b*, 4′ *b*, the distances that
their corresponding points in No. 2 are from the same assumed VP.
Join points 2′ and 4′ with 3′ by right lines, and parallel to them
draw two others indefinitely through point 3.   The plan of the line
c′c, No. 5, is next required.   This may be found in two ways : Either
by setting off from *xy* on the projector through 3′, the distance that
c in No. 2 is from the assumed VP, and drawing a line through the

MECHANICAL AND ENGINEERING DRAWING   131

points so found parallel to *xy* to cut those drawn through 3 in *c'c*,
No. 6 ; or by letting fall projectors from the ends *c' c* of the corresponding line in No. 5 to cut the same lines from 3 as before.   The
plans of the two short inclined edges of the section between *c'b'* and
cb in No. 5 are found by joining the corresponding points in No. 6 by
right lines.   On cross-lining the parts of the cube exposed by the
cutting plane in both views the required sectional plan and elevation
will be completed.

58. Now, to test the accuracy of these projections, the same reasoning holds good with reference to the sections thus found as with the
original uncut solid.   For it will be found, if the sections are correctly
projected, on applying the set-squares or a parallel-ruler to any one of
the bounding lines of the sections, the edge of that section on the
opposite face of a cube is exactly parallel to it.   This is as it should
be ; for if otherwise, the section obtained would be incorrect, for the
opposite faces of the cube being always parallel to each other, any plane
section of it will have parallel edges on opposite or parallel faces. ·

Before passing to the next problem, which is a variation of the last,
and somewhat more difficult of solution in that it involves a greater
amount of thought, it is necessary to point to the case of finding the
plan of the uncut cube, No. 4, Fig. 163.   The student will remember
the reference made in a previous problem to the change of position which
takes place in an original object when a plan of it is obtained from its
side elevation, as No. 4 is from No. 2 in Fig. 163.   In this case,
the matter of finding the plan is simplified by the position the cube
occupies with respect to the VP ; its vertical faces making, as they do,
equal angles with it, enable the corners figured 3, 4, 1 in No. 1 to be
used as the representatives of 2, 3, 4, in finding the plans of its edges,
shown in No. 4, by the lines 2, 2 ; 3, 3' ; 4, 4'.   Had the original
position of the cube, however, in No. 1 been any other with reference
to the VP than that actually occupied by it, then the plan of the cube
in No. 1 could not have been used to project from.   The reason of this
will be seen in the next problem, to which we now proceed.

Problem 61 (Fig. 164).—*Given the plan of a solid cube, with its
vertical faces making unequal angles with the VP ; to find its
projections, and sectional projections when the plane on which it
rests is inclined to the HP at 45°, and·the section or cutting plane
alters in direction.*

Assuming that the student has worked out the previous problem,
and repeated it, with the section plane in varied directions, he will be
the better able to arrive at a solution of the one now presented to him.
As the first part of it is nothing more than a repetition of the process
adopted in the last case, applied to the altered position of the original
object, no further assistance is given than that indicated by the construction lines shown, as the views almost explain themselves.   No. 1
is the plan of the original object or cube in the given position ; No. 2
its side elevation when inclined at the given angle ; No. 3 its front
elevation ; and No. 4 its plan at the same inclination.   Some of the

132   FIRST PRINCIPLES OF

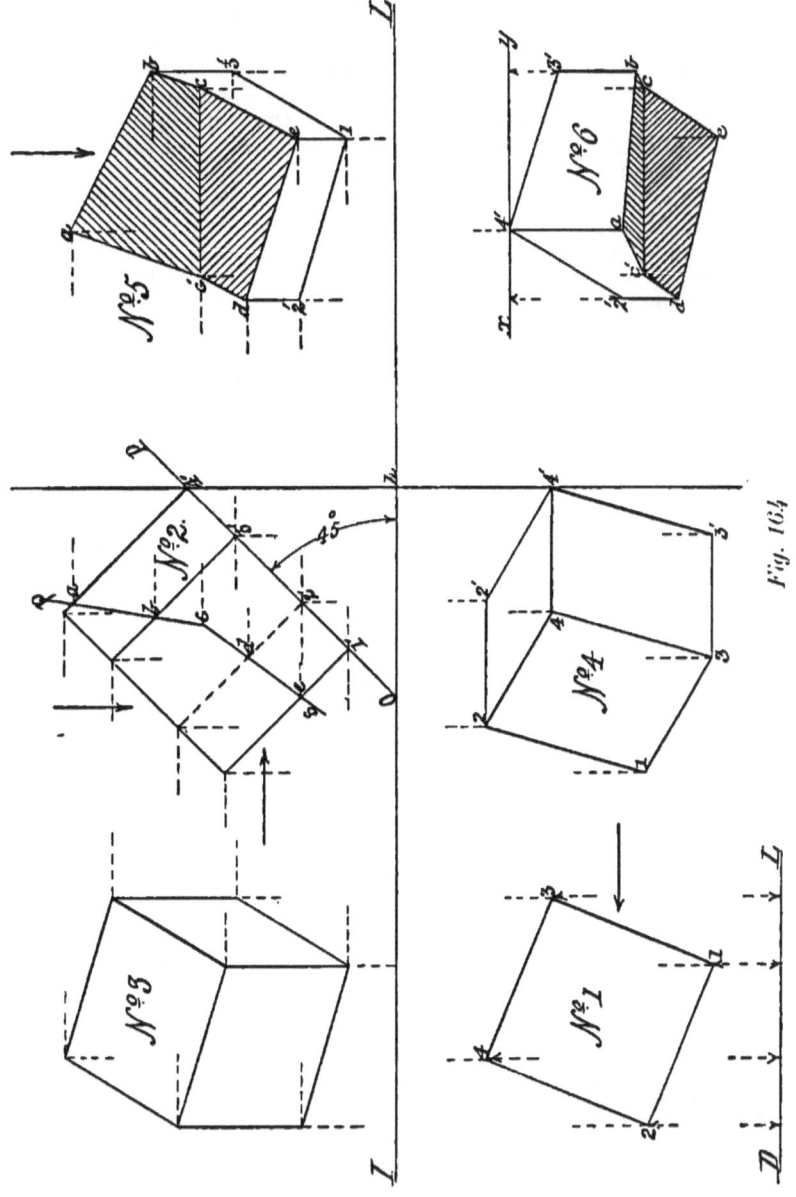

Fig. 164.

MECHANICAL AND ENGINEERING DRAWING   133

reference figures are purposely omitted that the student may feel less dependent upon such helps as he proceeds.

In views Nos. 5 and 6 (Fig. 164), the only point likely to cause difficulty in a first attempt at getting the sectional elevation and plan of the object will be at the *change* of the *plane of section* on the line $c'c$. This, however, should be thought out by the student, as it only involves the finding of the exact position of the two ends of this line on *opposite* faces of the cube. If it is borne in mind that it is the intersecting line of two plane surfaces, and that one of its ends is represented in elevation—c in No. 2—by a *point*, it is known at once that it must be a line passing through c, parallel to the HP, and therefore to the IL. Then, as it is parallel to the HP, its ends must be of equal height from the bottom face of the cube. This height is evidently the perpendicular distance between c and the plane OP; or the depth through the cube along the line $c'c$, square with its bottom face. With these data, the position of its ends should be easily determined without further guidance. Having found the elevation of this line, its plan offers no difficulty, as it may be got in one of two ways already explained.

As a test of the knowledge of our subject the draughtsman in embryo should by this time have acquired, we give for solution, without assistance other than that afforded by the few projectors shown in the views of the object, the following problem—

Problem 62 (Fig. 165).—*Given the plan of a hollow cube, with square holes, centrally situated in all its sides; to find its projections, when the plane on which it rests inclines 30° to the horizontal, and its sectional projections at the same inclination, when cut by a plane along a given line.*

The view No. 1 (Fig. 165) is the given plan of the cube, and the line sp in No. 2 the line of section. The different views should be drawn in on the sheet of paper in *the order of their numbers*, as they are consecutive projections, each one of which, when correctly found, enables that which follows to be more easily comprehended and drawn. This problem, with the one that follows it, may be looked upon by the young student as the *pons asinorum* in this particular part of the subject, which, when ably surmounted, may be regarded by him as one landmark passed on the road to his success as a practical draughtsman.

59. The problem referred to in the last paragraph is that of finding the projections of a skeleton cube, whose plan is given, resting on a plane inclined at an angle to both planes of projection.

In Fig. 166 is given the several views that would be obtained of such an object when in the assumed position, but without construction lines or any explanation of the process of finding its projections; the problem being intended as a test of the student's ability in applying the methods of procedure in its solution, which have previously been so fully dealt with. The order in which the several views should be drawn is the same as that shown in the three previous problems, Figs. 163, 164, and 165; and as the bounding edges of the different members of

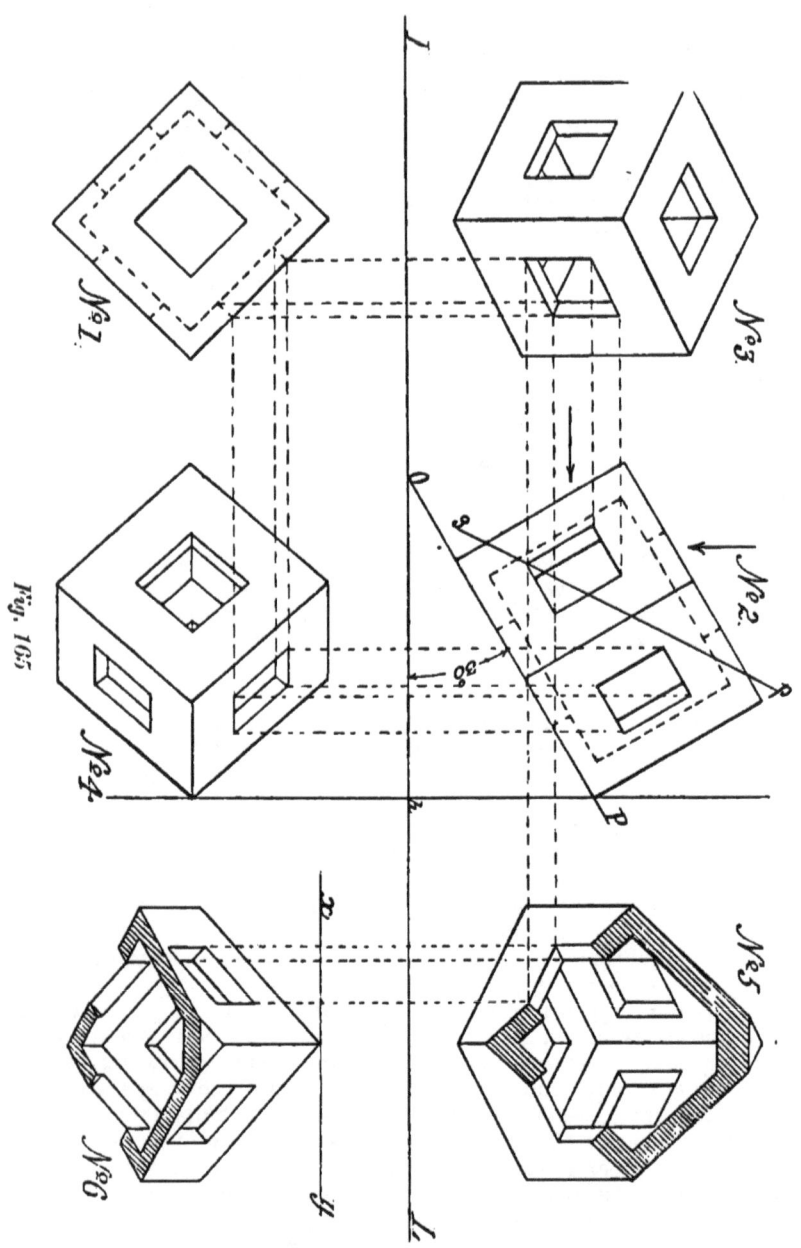

Fig. 165

MECHANICAL AND ENGINEERING DRAWING   135

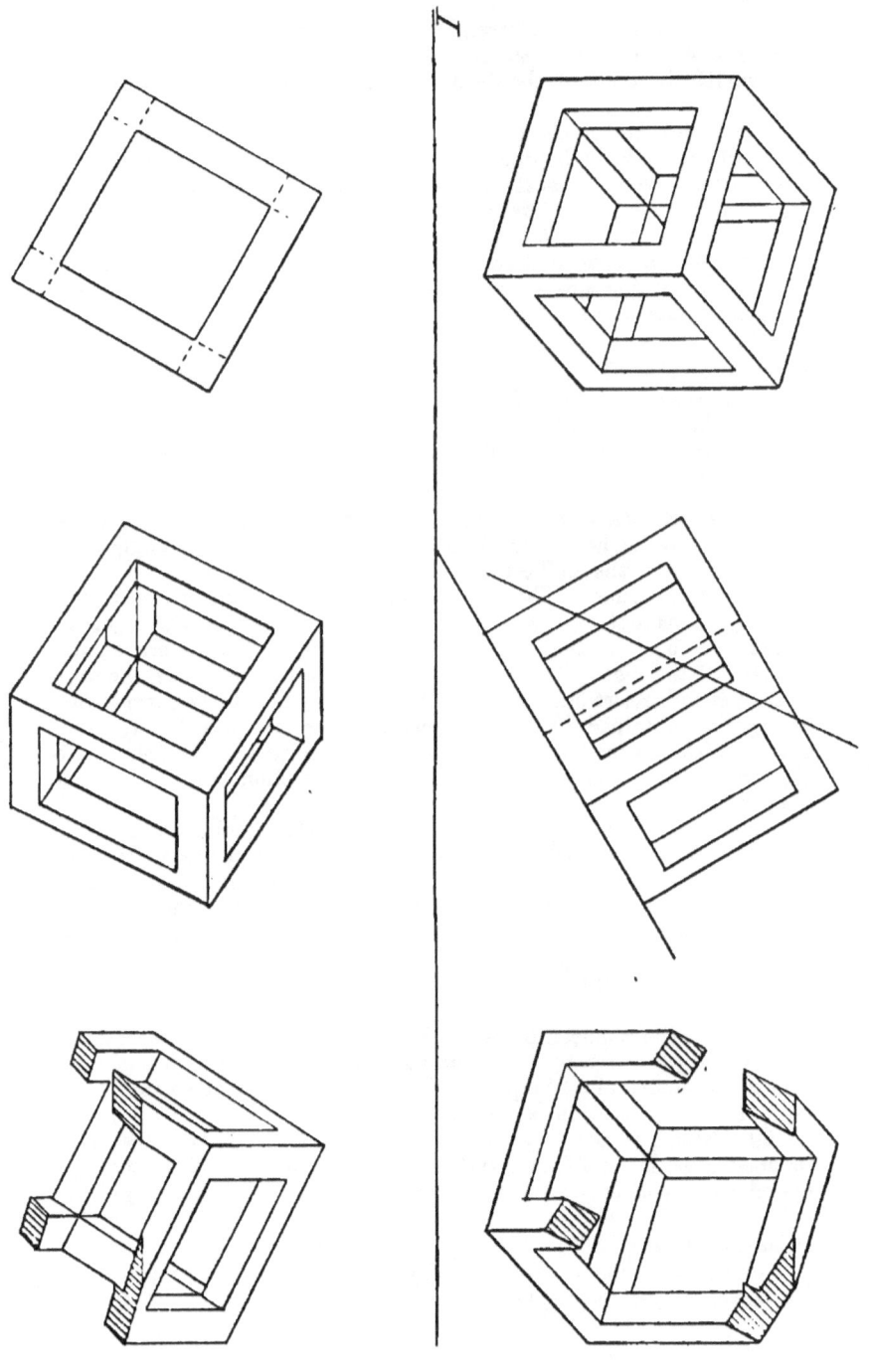

136   FIRST PRINCIPLES OF

the original object are all straight ones, and their surfaces plane, the
finding of its projections should offer no difficulty to the learner who
has attentively followed the development of the subject to its present
stage.

The projections of the prism and pyramid, or other plane-surfaced
solid, when similarly inclined, should also be found without trouble,
their surfaces being in most cases plane figures having straight bounding lines or edges. As, however, it is often necessary to obtain a view
of an inclined object, when turned through any given angle, to show
the method of procedure in such a case, the following problem is given,
as it combines the projection of both the prism and pyramid when
inclined to both planes of projection.

Problem 63 (Fig. 167).—*Given the plan No. 1, and elevation No. 2,
of a hexagonal pyramid, with a similar shaped prismatic base, to
find its projections; first, when its axis is inclined to the HP, and
parallel to the VP; and again when it is inclined to both the VP
and HP.*

If the inclination of the plane on which the given object is to rest
be, say, 30° to the horizontal, this at once determines the angle made
by its axis with the HP as one of 60°, the one angle being the complement of the other. Then the elevation No. 2 being given, to find
the projections first required, show as in No. 3 a view of the object on
a plane inclined to the HP at 30°. This is nothing more than a
transfer of the view given in No. 2, *directly* to this plane, as the same
faces of the object seen in No. 2 will be seen in No. 3, the former view
having been obtained by direct projection from its plan No. 1, *before*,
and not after, its inclination, as in previous problems. The plan of No.
3, given in No. 4, is found, as before, by projection from No. 3 and
No. 1.

Then to find the projections of the object when inclined to both the
VP and HP, as required in the second part of the problem, assuming
the angle its axis is to make with the VP is known, we proceed as
follows :—Draw in, in the HP, a line as *ab*, No. 5, making the same
angle with the IL that the axis of the solid is to make with the VP—
say one of 45°. To this line—which is a plan of the axis of the solid—
transfer the view obtained in No. 4, and from it, and No. 3, find by
projection that given in No. 6, then will the views Nos. 3, 4, 5 and 6,
Fig. 167, be the required projections of the original object, represented
in plan and elevation by Nos. 1 and 2 in the figure. In the same way
as here explained may the plan or elevation of any plane solid in any
desired inclined position be obtained.

60. Passing to the projection of solids bounded by curved surfaces
inclined to both the VP and HP, a similar combination of the cone and
cylinder is taken as the object for projection, the problem being—

Problem 64 (Fig. 168).—*Given the plan No. 1 and elevation No. 2
of a cone having a cylindrical base, its projections are required, first
when it rests on a plane inclined to the HP at 45°, with its axis*

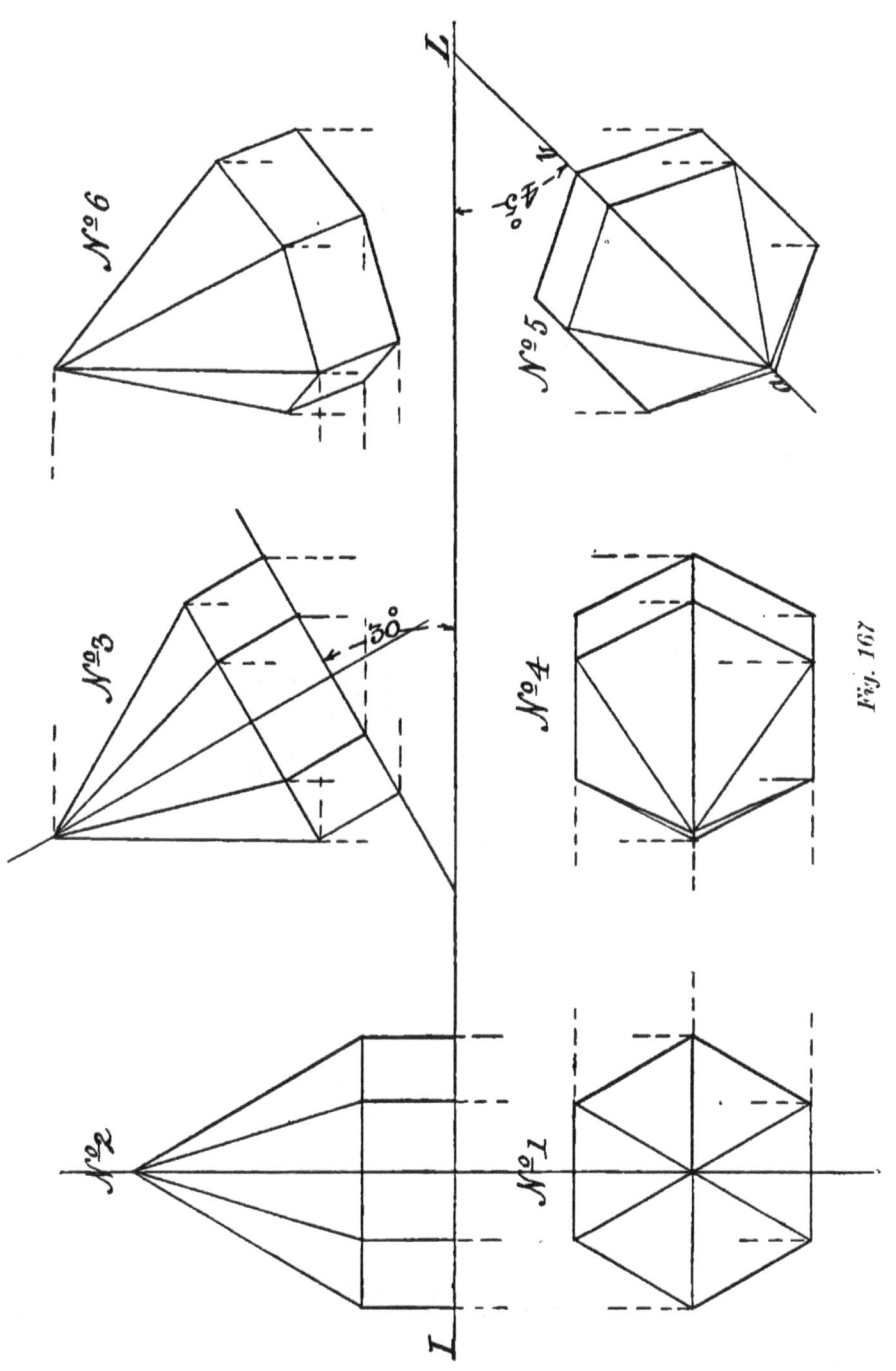

Fig. 167

FIRST PRINCIPLES OF

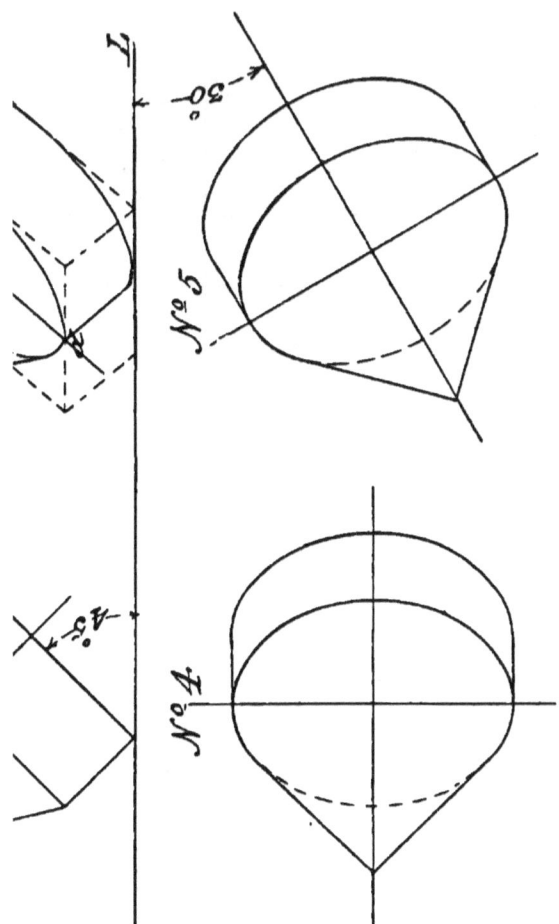

MECHANICAL AND ENGINEERING DRAWING   139

parallel to the VP, and after it has been swung from that position through a horizontal angle of 30°.

The method of procedure in this case being the same as in the previous problem, little difficulty should be found in obtaining the projections shown in Nos. 3, 4, 5 and 6 in the figure. As both portions of the object—the conical and the cylindrical—have a curved surface, greater care will be required in drawing in the ellipses, into which the top and bottom edges of the cylindrical portion are projected, but with this exception there is nothing to prevent correct projections of the object being easily found if the instructions previously given in reference to curved-surfaced solids are carefully followed.

To assist the student in determining the direction of the axes of the ellipses with respect to the IL, in the view required in No. 6 let him assume the base of the given object to be cut out of a square prism, as shown in dotted lines in No. 6, then the projection of the diametral lines a, b; c, d, of No. 1 on the upper face of this prism represented in No. 6 by the faint lines $a'b'$; $c'd'$, will be the axes of the ellipse sought, and a', b', c', d' will be four points through which the curve of the ellipse—into which the base of the cone is projected—has to be drawn. The lower edge of the cylindrical portion of the solid in its new position is, of course, projected into a similar ellipse as the upper edge, or base of the cone, but only half of it, as shown, will be seen.

As the projections of a sphere, in similar positions to the other solids taken as examples, offer little useful practice beyond that of the correct drawing of ellipses, we give as the concluding problem in this part of our subject, one in which the application of its principles is practically shown in the delineation of an ordinary "nut," in such positions as frequently occur in actual machine details. The problem is—

Problem 65 (Fig. 169).—*Given the plan and elevation of a six-sided chamfered nut, to find its projections when the plane surface on which it lies is inclined to the planes of its projection.*

Before proceeding to find the required projections of the nut, it may be noted in reference to such an article, that its flat *sides* at right angles to its top and bottom *faces* are called "panes"; that its height, or thickness, is generally equal to the diameter of the bolt it is intended to fit; its width across the "panes" or "flats" bears a certain fixed proportion to this diameter, and the "chamfer," or cutting down of its upper corners or angles, is usually at an angle of 45° with its top face.

To draw in the plan of a nut such as that shown at No. 1, Fig. 169, it is necessary to know first what its width across the flats is to be. This width for all nuts up to 6 in. diameter of bolt may be found in any engineer's pocket-book. With the given width—for the size of nut proposed to be drawn—as a diameter, describe a circle in the HP at a convenient distance from the IL, and from the same centre a second circle equal in diameter to that of the bolt for which the nut is intended. Through the common centre of these circles draw lines indefinitely, parallel and perpendicular to the IL. Then with the T-square and set-

140   FIRST PRINCIPLES OF

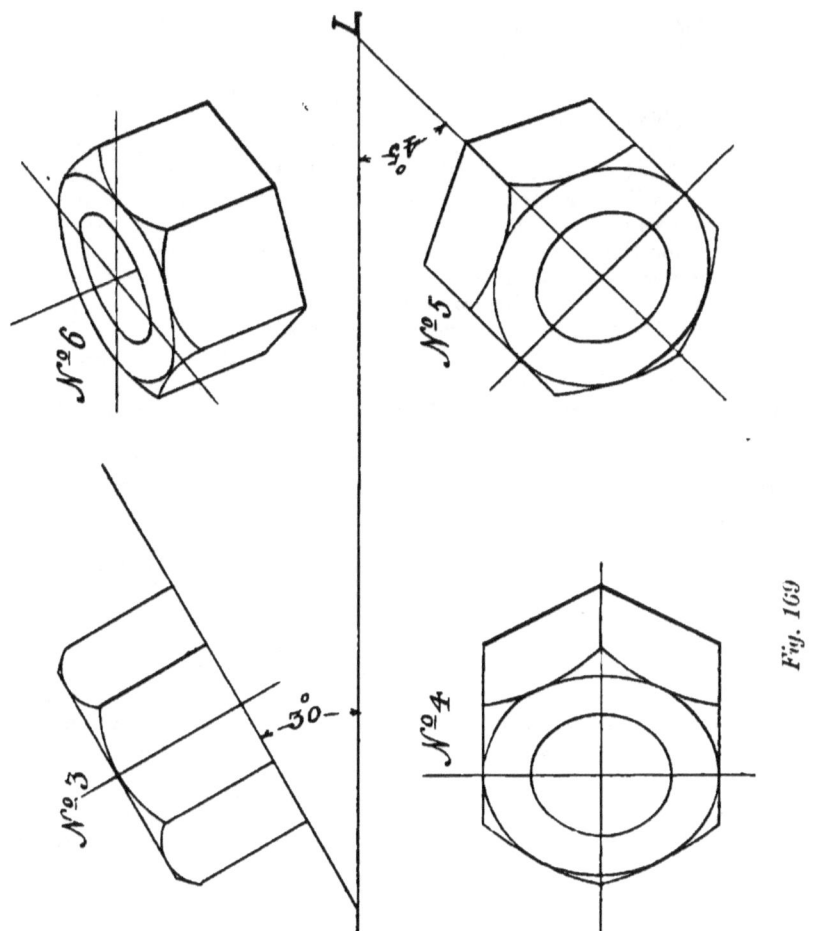

Fig. 169

MECHANICAL AND ENGINEERING DRAWING   141

square of 60°, draw lines tangential to the outer circle in the positions shown, and the resultant hexagon, with its inscribed circles, will be the plan of the intended nut when resting on the HP.

To find its elevation in this position, project over into the VP the four corner edges of the nut seen when looked at in the direction of the arrow in plan. These will, of course, be lines perpendicular to the IL. Now, as the *thickness*, or *height*, of an ordinary nut generally equals the diameter of its bolt, set off this diameter as a height—from the IL—and through it draw a line parallel to the IL. This line will represent the top face of the nut, and were it a plane-faced one and not chamfered, its elevation would be completed by making this line to cut the *four* perpendiculars previously drawn in. The chamfering is represented in elevation by the three circular arcs and the two short lines at an angle of 45°, tangent to the outer arcs, the reason for which will be fully explained when treating of screws, nuts, etc., later on.

The plan and elevation of the nut being thus given, its projections as required in the problem are now to be found. Assuming the angle of the plane surface on which the nut is supposed to rest to be one of 30° to the horizontal, draw in a line making with the IL such an angle. Transfer to this line the view of the nut given in No. 2, and obtain by projection its plan in this inclined position, as shown in No. 4. Then to find its elevation when swung through a horizontal angle—say of 45° —draw in its axial line in the HP making that angle with the IL, and transfer to it the plan of the nut found in No. 4, giving the view of it shown in No. 5. From this plan, and the elevation No. 3, find by projection the view given in No. 6, and the requirements of the given problem will be fulfilled.

An infinite variety of problems might be given in this very interesting part of our subject, but as they would only involve in their solution the correct application of principles which have been fully explained, we pass on to the elucidation of projection as applied to the "penetration" and "intersection" of solids, a knowledge of which is of the first importance to the would-be draughtsman.

CHAPTER XVII

THE PENETRATION AND INTERSECTION OF SOLIDS

61. UP to the present stage in our subject, all the objects chosen as examples to illustrate the application of the principles of projection have consisted of simple solids, with either plane or curved surfaces—or both—and have each been treated independently. They have not, of course, included *all* the elementary forms which, in combination, give shape to machine and engine details; but they have been such as will enable the student to delineate correctly any subsidiary solid—generally derived from one of the primary ones—which may enter into the construction of a complete mechanical structure.

As in the details of such a structure, some of the solids, which have so far been dealt with singly, are joined to, or made to penetrate, one another, their surfaces by such junction or penetration produce a line or lines which require accurate delineation in any drawing having pretensions to truthfulness. Such lines are, however, not only required for the correct representation of the objects in combination, but they are necessary to be known before the objects represented could be constructed in any material. It is then to the solution of problems—by the aid of projection—presented by such combinations, that attention is now to be directed.

In determining the lines of intersection of two solids, the simplest possible method of procedure—so long as it gives correct results—should be aimed at; and although plane-surfaced solids are probably not so often met with in combination in mechanical details as they are in building constructions, yet it is necessary that their intersections should be thoroughly understood by the draughtsman. The introductory problems in this part of the subject will therefore consist of those in which the intersections of plane-surfaced solids only are required to be found.

62. One of the first facts realized by the student in commencing the study of Projection, was that the intersection of two planes at *any* angle produced a *straight* line. It follows from this that the intersections of plane solids must also be straight lines, as the surfaces which intersect are planes. This consideration, then, is the key to the solution of any problem in which the intersections of plane-surfaced solids are

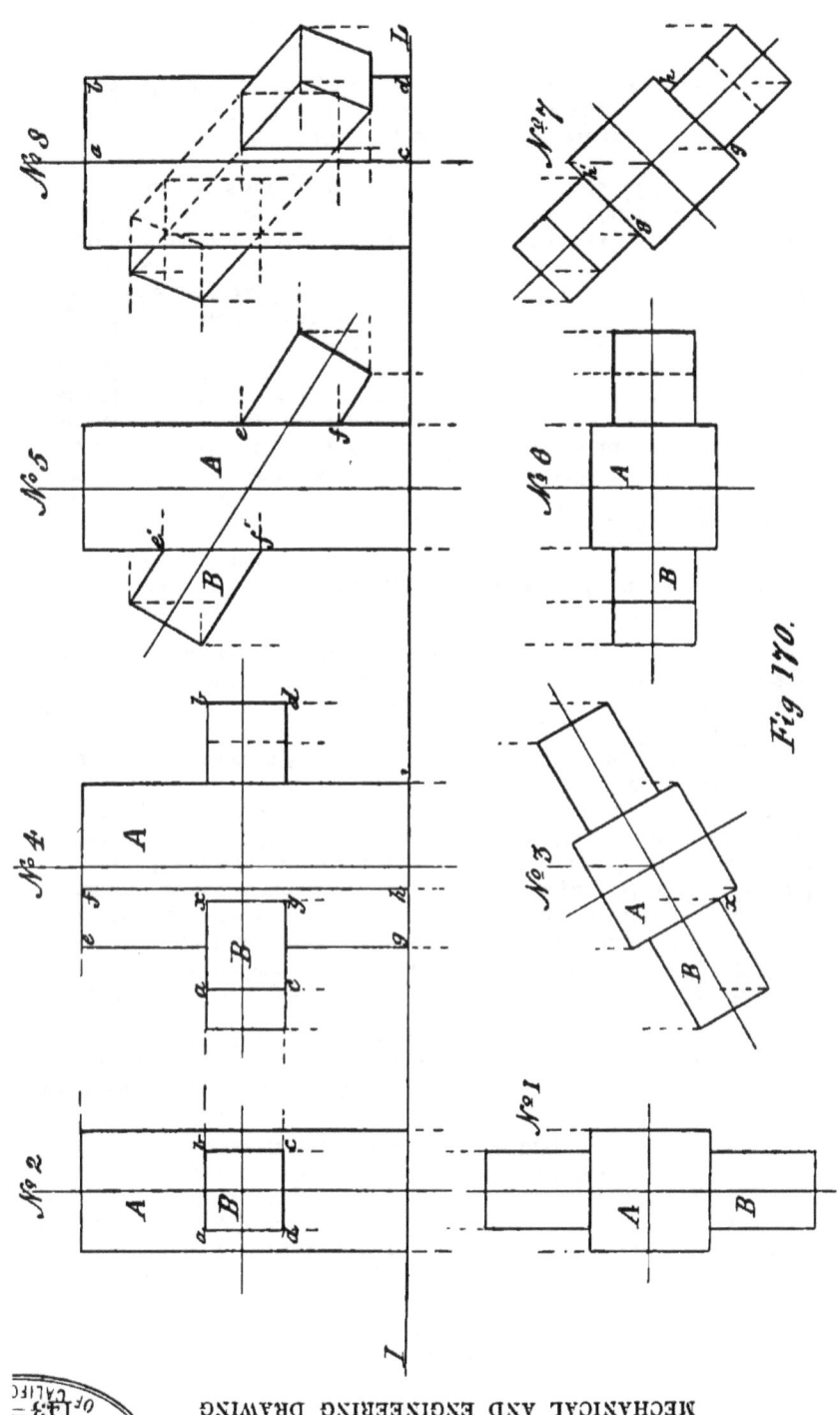

Fig 170.

144   FIRST PRINCIPLES OF

concerned ; for if through two such intersecting solids a section plane
is passed, the points where that plane cuts the surfaces of both solids
will be at once shown.   If, then, the two solids be imagined to be cut
simultaneously by a number of planes, and the points of intersection of
their surfaces by those planes be noted, a line drawn through those
points will be the "line of penetration" of one solid by the other.   As
the plane solids usually met with in machine details are prisms and
pyramids, with their frustums, the problems in this connection will deal
first with *prisms penetrated by prisms*.   The first is—

Problem 66 (Fig. 170).—*Given the plan No. 1, of two square prisms
A and B, of equal length, one penetrating the other at right angles ;
to find their elevation and lines of penetration.*

As shown in the plan, A is the *penetrated*, and B the *penetrating*
prism ; the former having two of its sides parallel to the VP, while all
four sides of the latter are perpendicular to it.   The lines of intersection
of the two, therefore, will only be seen on the *front* side of A, or that
nearest to the eye in elevation.   As that side is parallel to the VP, and
the prism B penetrates it at right angles, the lines of penetration will
coincide with those forming the front end of that prism.   Therefore,
find by projection the elevation of the prisms, in the positions shown in
plan, and the lines *ab, bc, cd, da,* in No. 2 will—although representing
the front end of prism B—be the lines of penetration sought.   In this
problem, as the four *sides* of the penetrating prism B are planes of
intersection with the two sides—or the one nearest to, and that farthest
from the VP—of the penetrated one A, the use of any section planes in
finding the lines of penetration is unnecessary.

*Next, let the two prisms be moved on the axis of A—as shown in No. 3—
until that of B makes an angle of 30° with the IL, and their elevation
and lines of penetration be required.*

Here, although the *axes* of the prisms are in the same relative position as before, that of B, being inclined to the VP, brings the vertical
sides of both at an angle with that plane, and will, in elevation, show
—as in No. 4—two sides of A and only one of B, and consequently but
one actual line of penetration, or that formed by the intersection of the
plane of the side *a, b, c, d* of the prism B, with the side *e, f, g, h* of the
prism A.   To show this line, draw a projector through point *x* in the
plan No. 3 into the VP, and it will cut the lines *a, b, c, d* in No. 4 in
x', y, which is the line sought.   The other lines of penetration of A by
B cannot be seen, as they are covered by those representing the front
edges of B, and the back ones do not come into view at all.

To find the lines of penetration of two intersecting square prisms
when the axis of one of them is inclined to *both* planes of projection,
the procedure is as shown in Nos. 5, 6, 7, and 8, Fig. 170.   First draw
in the elevation—as at No. 5—of the prisms with the axis of the penetrating one B, parallel to the VP, and at the angle it is to make with
the HP, and find their plan in this position.   Swing this plan, No. 6,
round on the axis of the prism A until that of B makes the required

MECHANICAL AND ENGINEERING DRAWING 145

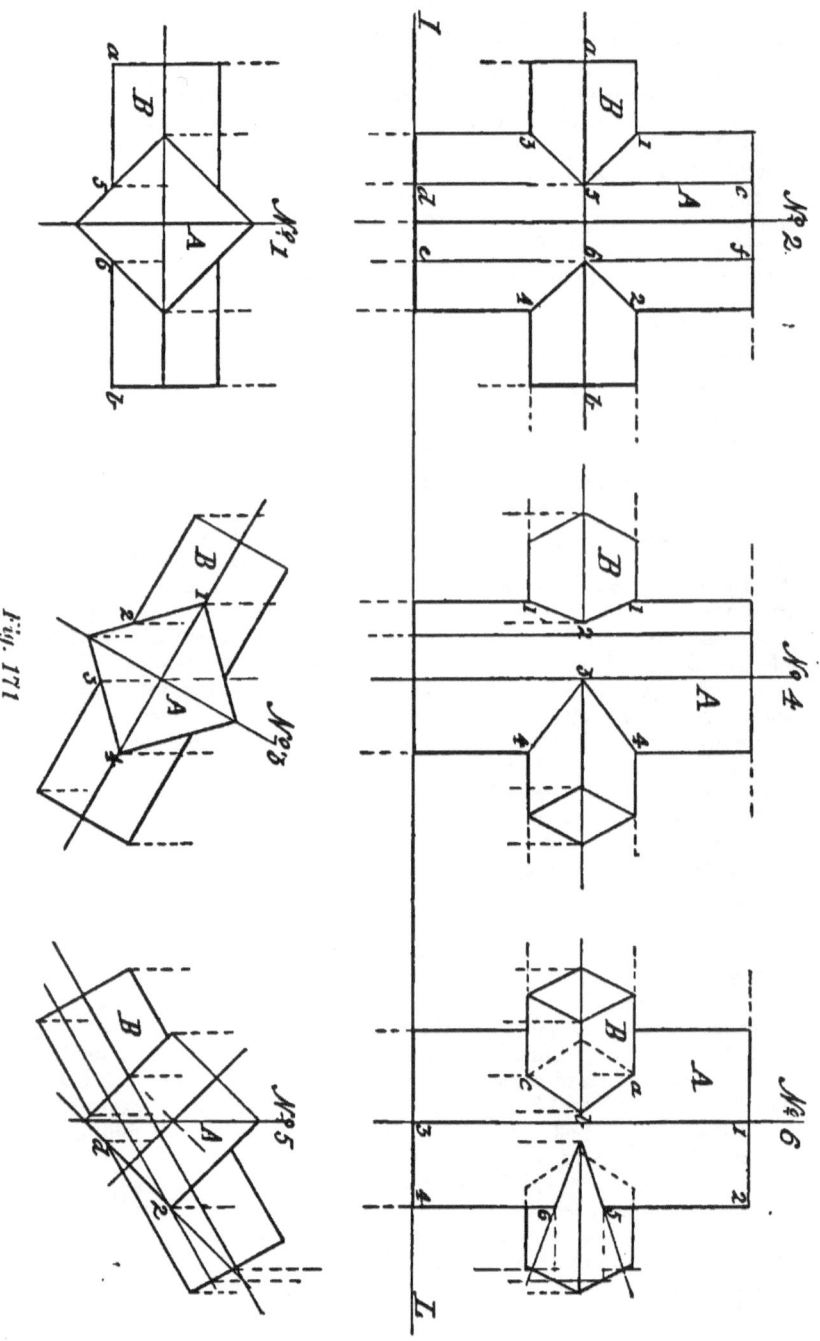

Fig. 171

FIRST PRINCIPLES OF

angle—say of 45°—with the VP or IL. Then with this plan, No. 7, and the elevation, No. 5, find by projection the elevation of the prisms in their new position, as shown in No. 8. Thus posed, but two lines of penetration will be seen—viz., where the prism B passes through the side *a*, *b*, *c*, *d* of A,—and are found by projectors drawn through *e* and *f* in No. 5, and *g* and *h* in No. 7. The lines of penetration of A by B on the side *opposite* to *a*, *b*, *c*, *d*—shown in dotted lines—are found by similar projectors from points *e'*, *f'* in No. 5, and *g'*, *h'* in No. 7.

The next problem is a case in which a diagonal of the intersected prism is in the same plane as that of a diagonal of the prism intersecting it.

Problem 67 (Fig. 171).—*Given the plan of two square prisms of equal length, penetrating each other at right angles, their axes and a diagonal of each being in one and the same plane, to find their lines of penetration.*

First, let the axes of both prisms be parallel to the VP, then No. 1 (Fig. 171) will be their plan. From this, first find the elevation of the prism A, making it equal in *height* to the length—given in the plan— of B. At the middle of this height draw in the axial line of B, and on it project over from the plan No. 1, the two *ends* of that prism. Now the lines of penetration of A by B may be found in two ways, either by projecting over to the elevation the points of intersection of the visible edges of B with A, seen in plan, or by the use of section planes. By the first method it will be seen at a glance, that in elevation only two sides with their three edges of the prism B will be seen as penetrating A. These are the top, bottom, and front edges, the two former of which intersect the right and left vertical edges of A in points 1, 2, 3, 4. To find where the front edge of B enters and leaves A, project over from the plan of this edge the line *ab* No. 1, to its elevation *ab* No. 3, the points 5, 6; join points 1, 5; 5, 3; 2, 6; 6, 4 by straight lines, and they will be the lines of penetration sought.

63. To find the same lines by the use of intersecting planes, it will be seen at once that if a plane be caused to pass through the axes of both prisms parallel to the VP, it would intersect the prisms in the points 1, 2, 3, 4, No. 2; and if a second plane, parallel to the first and tangent to the edge *ab* of the prism B, be passed through A, it would intersect the edge *ab*, No. 1, in the points 5 and 6, as the section of A made by this plane would be the parallelogram *cdef* shown in faint lines in No. 2, the points of intersection of the edge *ab* by it being 5 and 6, which, joined to 1 and 3 and 2 and 4, give the same lines of penetration as by the first method. It will be seen, on studying this figure, that the resultant lines of penetration are nothing more nor less than the intersections of *pairs of planes at an angle*, and it is in the judicious application of such section planes that the whole art of finding the lines of penetration of solids consists. It will, however, be evident that in the problem just solved, and that which follows it, their use in practice would be dispensed with, as the lines sought can be found at once by simple projection from the plans of the objects; but, as it is advisable

MECHANICAL AND ENGINEERING DRAWING   147

that at this stage the student should understand the application and use of such planes, they have been introduced thus early.

For our next problem, the same solids are taken as in the last, but instead of both axes being parallel to the VP, that of the prism B is inclined at an angle to it. This position is shown in plan in No. 3, Fig. 171—a transfer of No. 1 at an angle to the VP or IL—and it is required to find the elevation of the prisms, and their lines of penetration when so posed.

To do this, first find by projection from No. 3 and No. 2, the elevation of the prisms as if they were entire. From the plan No. 3, draw projectors from the points 1, 2, 3, 4, in it, where the edges of the prism B—that will be seen in elevation—intersect those of A; then straight lines joining these points will be the lines of penetration sought, as shown in No. 4.

64. In cases where the axes of the intersecting solids are not in the same plane, it may happen that the lines of penetration cannot be found by direct projection alone, as in the previous problems, but will necessitate the use of section planes to discover them. As an instance of this, take the following problem—

Problem 68 (Fig. 171).—*Given the plan of two square prisms of equal length, intersecting each other at right angles, but having their axes in different planes ; to find their lines of penetration.*

Let No. 5, Fig. 171, be the given plan of the prisms, the axis of A being vertical and that of B horizontal, but inclined to the VP. Obtain by projection the elevation of the prisms as if they were entire. The two lines *ab*, *bc* of penetration, of A by the prism B, that will be seen in elevation, assuming it to enter A from the *left*, can be found by direct projection from the plan, as in the previous problem. Those through the front side of A to the right may be found in either of two ways, by the use of a section plane.

First assume a plane passing vertically through the prism B, and coinciding with the right-hand front side of prism A, a plan of which would be the faint line *xx*, No. 5. The section of B by this plane would be a triangle, its apex being the point *d* in the frontmost edge of the prism B, and its base, a line across the *right* end of that prism, perpendicular to the HP. Find, by projection, the elevation of this triangular section, as shown in No. 6, and the points 5, 6 in its sides will be those where the two front sides of the penetrating prism B intersect the vertical line 2 4, or the right edge of the prism A in passing through it. Right lines joining points 5 and 6 with *d* the apex of the triangle thus found, will be the lines of penetration required.

The same lines may be found by assuming a plane to pass vertically through *both* prisms parallel to the axis of B at point 2 in plan ; or through the line 2 4 in elevation. Find, by projection, the elevation of the rectangular section of B made by this plane, and it will give the points—5 and 6—in No. 6, as those where the two front sides of the prism B intersect the vertical edge 2 4 of prism A. These points, joined

148   FIRST PRINCIPLES OF

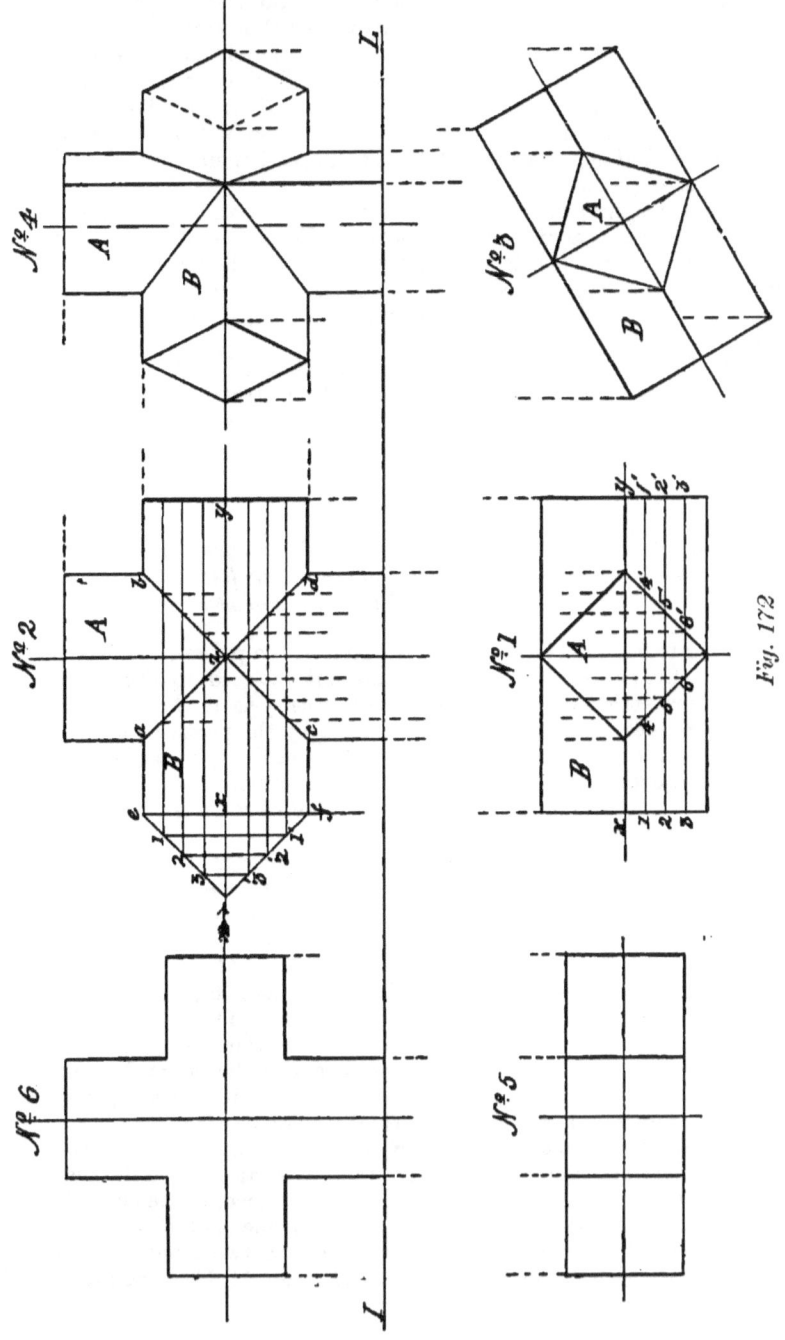

Fig. 172

MECHANICAL AND ENGINEERING DRAWING   149

by right lines to the vertical projection of point d in the plan, give the same lines of penetration as before obtained.

65. Had the prisms in the foregoing problems been *equal* pairs, the resultant lines of penetration would of course have been different. In such case, when seen in elevation as in No. 6, Fig. 172, there would be no lines of penetration at all, as the two visibles faces are in one and the same plane, and therefore show no juncture. If, however, their sides were inclined to the planes of projection—their axes being still at right angles and in one plane parallel to the VP—then the lines of penetration would be as shown in No. 2, Fig. 172, taking the form of a cross, due to their surfaces intersecting each other equally. This may be proved to exactness by the use of section planes. Let No. 1, Fig. 172, be the plan of two equal-sized prisms A and B, intersecting each other at right angles, and with their axes in one and the same plane, and let it be required to find their lines of penetration.

From the position, shown in plan, in which the prisms are with respect to each other, it is evident that a section plane passing vertically through *both* their axes, would give equal rectangular sections of each, and four points of intersection with the prism B; a and b being those of entering, and c and d those of leaving it. Then as the prisms are equal, it follows that any number of parallel planes passing simultaneously through both of them would give equal sections and points of intersection; lines, therefore, drawn through these points, taken in order, will give the lines of penetration of one solid by the other.

To find these lines in the given problem, obtain by projection from the plan No. 1, an elevation, as in No. 2, of the prisms as if they were entire. Parallel to the axis of B in No. 1, draw the lines 1 1', 2 2', 3 3', which will be the edge view, or plan, of the section planes to be used. Find the elevation of each of the sections of the prism A, produced by these planes, by vertical projectors from the points 4 4', 5 5', 6 6', in No. 1; then as the prisms are equal, and the section planes are assumed to pass through both, the horizontal one B will give similar sections to that of A. To find their bounding lines, produce the axial line $x y$ of the prism B in No. 2 to the left, and on it draw in a *half-end view* of that prism looked at in the direction of the arrow on the left. On this set off from the line ef the distances that the lines 1 1', 2 2', 3 3', in plan No. 1 are from the axis of B, and through the points thus found draw vertical lines to meet the two sides of the triangles in the points 1 1', 2 2', 3 3'. Then parallel to the IL, and through these last-found points, draw lines from end to end of the prism B, as shown in No. 2. The points where these lines cut the corresponding vertical ones drawn through points 4 4', 5 5', 6 6', in No. 1, will be points in the lines of penetration of the two prisms. On drawing lines through these points, they will be found to be straight ones, crossing each other at the point z, where the two frontmost edges of the prisms intersect. Although these lines might have been at once obtained by joining the points $a d$, $b c$, in No. 2, by straight lines, it is advisable that the student should be able to prove that they become straight, rather than take the matter for granted.

If the axis of one of such equal prisms be inclined to the VP,

150   FIRST PRINCIPLES OF

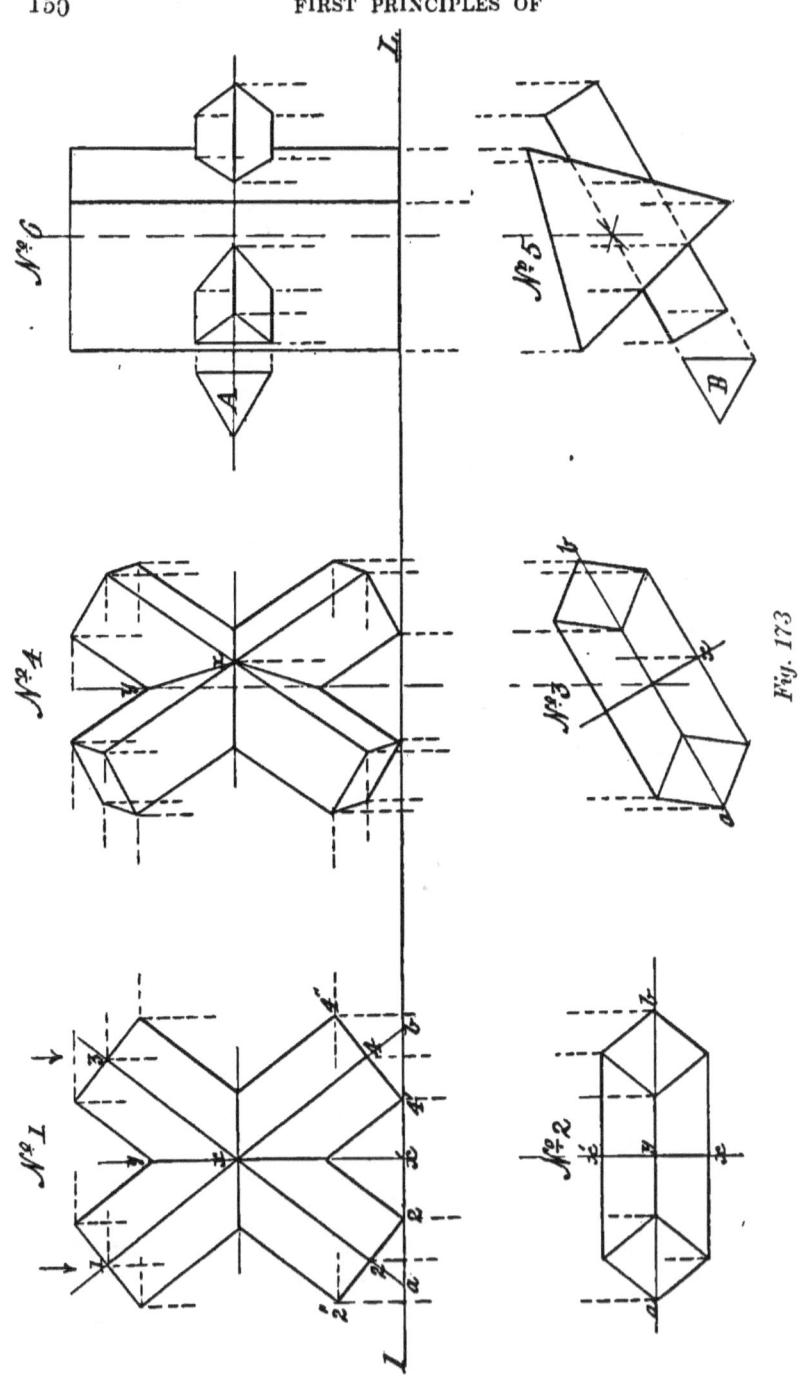

Fig. 173

MECHANICAL AND ENGINEERING DRAWING   151

the other remaining parallel to it as before, then the lines of penetration will be as shown in No. 4; obtained by first finding the elevation of the prisms as if they were entire, and then projecting over from No. 3—which is No. 1 with the axis of the prism B at the desired inclination to the VP—to the corresponding edges in No. 4, the points where the edges of B that will be seen in elevation intersect those of the prism A. The right lines joining these points as shown in No. 4 are then the visible lines of penetration of the two prisms.

For the next problem a case is taken where, although the axes of both prisms are still in the same plane, yet neither of them is vertical or horizontal.

Problem 69 (Fig. 173).—*Two equal square prisms have their axes in one and the same plane, and intersect each other at the middle of their length; give the elevation and plan of the prisms, and their lines of intersection when their axes are parallel to the VP, but inclined to the HP at $52\frac{1}{2}°$.*

As there would be no visible lines of penetration of the two prisms given in the problem, if their sides were parallel and perpendicular to the plane in which their axes lie, it is assumed that that plane divides each square prism into two equal triangular ones, and as their axes are inclined to the HP, it is evident that their elevations must first be drawn. Now the axes being in a plane which is parallel to the V P, to draw the elevation of the prisms in the position stated in the problem, proceed as follows :—

Draw in, in the VP, two lines at an angle of $52\frac{1}{2}°$ with the IL, intersecting each other. This may be done by using a " scale of chords " or a " protractor," or with the set-squares of 45° and 60°. If neither of the former is to hand, take the set-square of 45°, and at a convenient point—say a—in the IL, draw a faint line at 45° with it; through the same point draw a similar line with the 60° set-square; bisect the angle formed by these lines, and the line of bisection will be at an angle of $52\frac{1}{2}°$ with the IL, as it is the mean betwen 45° and 60°; for $\frac{45 + 60}{2} = 52\frac{1}{2}$. Through a convenient point—as x—in this line, let fall a perpendicular to the IL, cutting it in point x'; make $x' b$ in that line equal to $x' a$, and from b through x draw a line indefinitely. Then as the lines passing through x each make an angle of $52\frac{1}{2}°$ with the IL, they will represent the axes of the required prisms, and from their position with respect to the VP, their frontmost edges also. Set off from x, in points 1, 2, 3, 4, half the length of the intended prisms, and at those points draw lines at right angles to the axial ones; those drawn through 2 and 4, cut the IL at $2'$, $4'$; in these make 2 2'', 4 4'' respectively, equal to 2 2', 4 4'. Through 2' 2'' and 4' 4'' draw lines parallel to the axial lines 2 x 3, and 4 x 1, and the resultant rectangles will be the elevations of the prisms as if they were entire, resting on the HP, on the two opposite corners, 2', 4', of their lower ends.

Having drawn in the elevation, the plan of the prisms may now be found. Their axes being parallel to the VP, and in one plane, that

152   MECHANICAL AND ENGINEERING DRAWING

plane will be correctly represented by a line—as ab in No. 2—parallel
to the IL. It is evident such a plane will pass through the two
diagonals of the ends of the prisms—seen from above in the direction
of the arrows—which are parallel to the VP. Having found the plan
as shown in No. 2, the lines of penetration of the prisms are now required.
These in elevation will be the intersections at equal angles of the front
sides of the prisms. The points of intersection of the edges of the prisms
are those between which the lines of penetration will pass, and as the
prisms are equally inclined to each other, those lines will pass through
point x in No. 1 at right angles to each other, as shown. In the plan
No. 2, only two such lines will be seen—viz., xy, and the one directly
behind it in elevation—but which, on account of their position—both
being in one plane—merge into one line xyx', shown in No. 2. Now,
as such junctions of prisms as are shown in No. 1 (Fig. 173) often
occur with their axes in a plane at an angle with the VP, we give such
an instance in the following problem—

Problem 70 (Fig. 173).—*Given the plan of the equal square prisms
with their axes inclined to each other at the same angle as in the
last problem; to find their elevation and lines of penetration, when the
plane of their axes is inclined to the VP at an angle of 30°.*

The position the prisms will now assume will be tantamount to
swinging the two together as shown in No. 1 (Fig. 173), on the corner
figured 2′, of the left-hand one, through a horizontal angle of 30°. If
then the plan of the prisms obtained in No. 2 be drawn in with the
line ab, making that angle with the IL, as shown in No. 3, then the
elevation of the prisms in this position may be obtained by direct
projection from No. 3 and No. 1, care being taken that the projectors
used in doing so are drawn from *corresponding* points in each. If due
consideration be given to the relative position of the edges of the prisms
to each other, no difficulty should be experienced in obtaining a correct
projection of them, as shown in No. 4. The lines of penetration in
elevation do not, of course, in this case *show* as being at right angles to
each other, although they are virtually in that position.

Nos. 5 and 6 in Fig. 173 show the application of the same method
of procedure to the case of the penetration of triangular prisms as
to those of square section; which require no further explanation than
that afforded by the projectors shown in the diagram. One side of the
penetrating prism is shown as coinciding with and passing through the
axis of the penetrated one. In each case the cross-section of the prism
is an equilateral triangle, the size of the penetrating one being shown by
the end views given at A and B. As a multiplicity of similar examples
to the foregoing would only show the application of the same principles
in determining the lines of penetration of prisms by prisms, a few
problems will now be taken in which the intersections of pyramids
are involved.

CHAPTER XVIII

THE INTERSECTIONS OF PLANE SOLIDS (*continued*).

66. BEFORE passing directly to the finding of the lines of penetration of one *pyramid* by another, the solution of such a problem will come much easier to the student if some preliminary practice is had with the case of a pyramid penetrating a *prism*. As such a combination is often used in practice, in giving form to simple ventilators, cowls, etc., we give as problems in this connection those in which the two solids are combined for such purposes.

The first combination, shown in plan and elevation in No. 1 and No. 2 (Fig. 174), is the simplest possible, it being that of a square prism, with its axis vertical and sides parallel and perpendicular to the VP, penetrated by a square pyramid, the axis of which is parallel to both the VP and HP, and in the same plane as that of the prism.

From the relative position of the two solids, as shown in the figure, it is evident that neither in plan nor elevation will any lines of penetration be visible in this case, as the *two* faces *a* and *b* of the prism, which are penetrated by all *four* faces of the pyramid, are in plan and elevation—through being perpendicular to both planes of projection—each represented by *one* straight line only, with which the actual lines of penetration coincide, and therefore cannot be seen. Had the faces of the prism, however, been in the least *inclined* to the VP—still remaining perpendicular to the HP,—then one line of penetration only would have come into view, dependent upon the inclination of the axis of the pyramid to the VP.

For all purposes where combinations of solids are used to give form to mechanical details, it is usual in practice to so combine them that their intersections shall be symmetrical. That is to say, that when the *form* of the article to be made requires that one solid should penetrate another, the penetration is made in such a way that an *axial* plane passed through the solids shall divide *each* of them into *equal* parts. This, although it necessitates the axes of both solids being in one and the same plane, does not necessarily prevent either the axes or sides of the solids from being *inclined* to the VP or HP. As instances we may take the following cases put in the form of problems—

153

154   FIRST PRINCIPLES OF

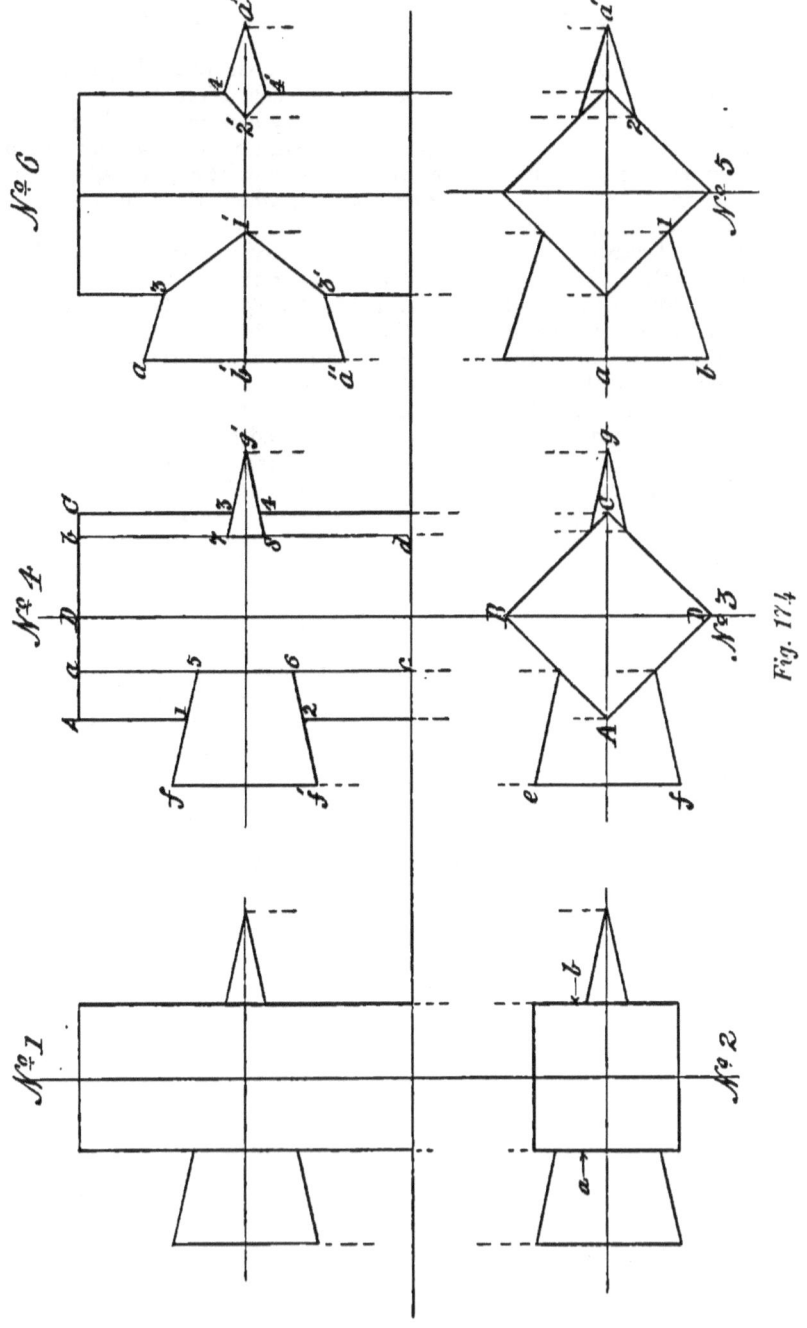

Fig. 174

MECHANICAL AND ENGINEERING DRAWING   155

Problem 71 (Fig. 174).—*Given the plan of a square prism, with its axis vertical and its sides equally inclined to the VP, penetrated by a square pyramid having its axis parallel, and its opposite sides inclined to one of the planes of projection and perpendicular to the other; to find the elevation and lines of penetration of two solids.*

Let No. 3 (Fig. 174) be the plan of the two intersecting solids. Then from the definition of a prism, and of a pyramid, it will at once be manifest that an axial section—or one along the line A C y—of these solids will in elevation give a rectangle for the prism, and a triangle for the pyramid. Obtain these sections by projection from the plan; and the points 1, 2, 3, 4 in that of the pyramid will be the limiting ones in the penetration. Complete the elevation of the prism by drawing in its front vertical edge. To determine the intersections of the front side of the pyramid $fg'f'$, No. 4, with the two inclined vertical sides of the prism seen in elevation, find by projection the elevation of the lines formed by the meeting of those sides. Now the points a and b, in the line fg in No. 3, are evidently the *plans* of the lines required. Therefore, through them draw projectors into the VP, and they will cross the front face of the pyramid between the points 5, 6, and 7, 8 in its top and bottom edges, and will give the lines of penetration sought. These being put in, in full, and joined up to f and f', and g' respectively, will complete the solution of the problem.

As shown in a previous problem, the lines of penetration just found can also be determined by assuming a section plane to pass vertically through the prism on the line of the front face of the pyramid giving the rectangular section—$a\,b\,c\,d$, No. 4—shown in faint lines, intersecting the triangular section first found in the points 5, 6, 7, 8; and thereby giving the lines of penetration sought.

The second case of the penetration of a prism by a pyramid is that in which the prism is in the same position with respect to the VP as before; but the pyramid penetrates it in such a way that an axial plane passing through both, divides each solid into two equal ones having triangular bases.

Problem 72 (Fig. 174).—*A square prism, in the same position as in the last problem, is penetrated by a square pyramid having its sides equally inclined to both planes of projection; required their plan and elevation and lines of penetration.*

With one exception, the plan of the solids in the positions given in the problem will be similar to that given in No. 3 (Fig. 174). The exception is caused by the altered position of the pyramid. The triangle efg in No. 3 represented a *flat* surface perpendicular to the VP—one of the faces of the pyramid—whereas in No. 5, the pyramid having been turned on its axis through 90°, has brought two of its sides into view in both plan and elevation—equally inclined to the planes of their projection, and therefore requiring a third line to represent

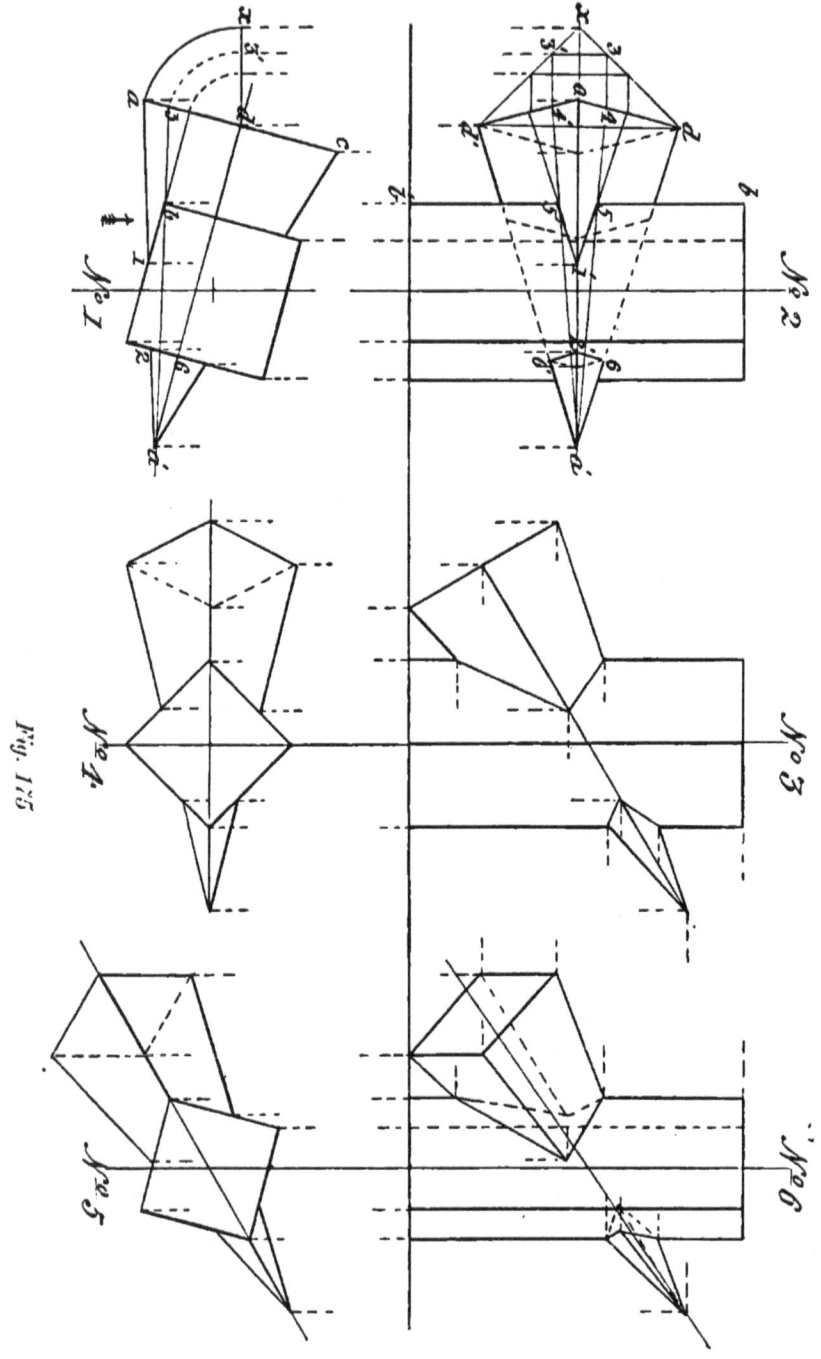

Fig. 125

MECHANICAL AND ENGINEERING DRAWING   157

their junction.   This line or edge in No. 5 will be $a\,a'$; that part of it
where it passes through the prism being out of sight.   From this plan
No. 5, find by projection the elevation No. 6 of the solids as if they
were entire.   For the lines of penetration that will be visible, draw
projectors from the points 1, 2, in the line $b\,a'$ in No. 5, to cut $b'\,a''$ in
No. 6 in the points 1', 2' ; then lines drawn from these last-found points
to 3, 3' ; 4, 4' ; or where the upper and lower edges of the pyramid
enter and leave the prism, will be the lines of penetration required.

67. Although, as stated in a previous paragraph, it is usual in
practice to make the intersections of solids in the design of mechanical
details symmetrical, cases may sometimes arise in which this symmetry
cannot be adhered to.   Through the exigencies of *position* of the solid
penetrated, it may be impossible to bring the axes of both into one and
the same plane, although at right angles to each other.   To meet such
a case the following problem will show the method of procedure in
determining the lines of penetration.

Problem 73 (Fig. 175).—*Given the plan of a square prism, penetrated by a square pyramid, their axes being in different planes ; required the elevation and lines of penetration of the solids.*

From No. 1 (Fig. 175), which is the given plan of the solids, it is
seen that the axis d, a' of the pyramid passes through the prism in a
direction parallel to two of its opposite sides, but some distance in
front of its axis.   To find the lines of penetration when in such a position, obtain by projection the elevation of the solids as if they were
entire, both of them being of the same dimensions as in the previous
problems.   As a vertical plane passed through the axis of the pyramid
would divide it into two equal triangular ones, and give an isosceles
triangle for the section made by the cutting plane, it is evident that
any vertical plane passing through the apex of the pyramid would give
a similar—though not equal—triangular section.

Now, in looking in the direction of the arrow in No. 1, it is seen
that the frontmost edge $a'a$ of the pyramid enters the *left* visible face
of the prism at point 1, and leaves it *on* the *right* face at point 2.
These points projected over to the corresponding edge of the pyramid
in No. 2, give 1', 2''.   To find where the vertical edge $b\,b'$ of the prism
—seen in elevation in No. 2—enters and leaves the two front faces of
the pyramid, draw (in No. 1) a line through a', its apex, and the
corner b of the prism, and produce it to cut the base line $a\,c$ of the
pyramid in the point 3.   We have then to find by projection the elevation of the section thus made, to enable us to obtain the points required.   Taking this line, $3\,b\,a'$, as the edge view, or plan, of a vertical plane passed through the pyramid, it will give a triangle for the
section.   To determine it we must first get an end view (in elevation)
of the base of the *front* half of the pyramid.   Now, its whole base is a
square, with diagonals equal in length to the line $c\,d\,a$—in No. 1—one
of them being vertical and the other horizontal.   To draw the half-
base required, find the elevation of the vertical diagonal of the base of
the pyramid, represented in No. 2, by the faint line drawn between the

158   FIRST PRINCIPLES OF

points d and d', they being the top and bottom ends of that line; produce the axis $a\,a'$ of the pyramid, and on it draw the half-square $d\,x\,d'$ as shown.

Now the assumed section plane, represented by the line 3 $b\,a'$ in No. 1, cuts through the base of the pyramid at the distance d 3 from its vertical diagonal. To find the line on the half-base so made, through d in No. 1 draw a line indefinitely, parallel to the IL; with d as centre and radius $d\,a$, draw an arc, cutting the line through d produced, in the point x, and from the same centre, with a radius equal to d 3, describe an arc cutting $x\,d$ in 3'. As the line $x\,d$, No. 1, is a plan of the half-square $d\,x\,d'$, No. 2, swung round on its diagonal d, d' as a hinge, until it is parallel to the IL, on this half-square—or front half-base of the pyramid—can now be found the exact length of the base of the isosceles triangle produced by the cutting plane, and from it the section itself, and the lines of intersection of the sides of the prism and pyramid.

For the triangular section, draw a projector into the VP through point 3', in the line $x\,d$, No. 1, and it will cut the lines $x\,d$ and $x\,d'$ in No. 2, in points 3, 3'; the distance between which is the length of the base of the triangular section. For its *sides*, cut by projectors drawn through points 3, 3', parallel to the IL, the lines $d\,a$, $a\,d'$, in points 4, 4', and through these and the apex a' draw *faint* lines as shown in No. 2; then the points 5, 5', where these faint lines—or edges of the triangular section—cut the vertical edge $b\,b'$ of the prism, are two of the points sought. Join 1' in the line $a\,a'$ to these points by straight lines, and they will be the lines of penetration on the *left* front face of the prism. For those on its *right* face, find by projection the elevation of the points—2 and 6—in the edges of the pyramid where they leave that face, and join these as before by the straight lines 6 2', 2' 6', and they will be the lines of penetration seen on that face. For the corresponding lines on the *back* faces of the prism—not seen, but shown dotted—the same method of determining them is employed as for those that are directly in view.

68. As an example of the case of the penetration of a prism by a pyramid, when the axes of the two solids are in one plane, but *not* at right angles to each other, we give in Nos. 3, 4, 5, 6 (Fig. 175) the resultant projections and lines of penetration of the solids when having these relative positions. Nos. 3 and 4 show the plan and elevation of the solids when the plane of their axes is parallel to the VP, and therefore perpendicular to the HP; and Nos. 5 and 6, those of the same solids when that plane is inclined to the VP, but still in the vertical position. As the required projections and lines of penetration should now be found by the student without further assistance than that given—by the projectors—in the diagram, we pass on to the determination of the lines of penetration of one pyramid by another.

The first case is that of the penetration of a square pyramid by a similar one of equal size, the axes of both being in one plane, parallel to the VP, and at right angles to each other.

First, let the penetrated solid be in such a position that the *diagonals* of its *base* are respectively parallel and perpendicular to

MECHANICAL AND ENGINEERING DRAWING   159

the VP, and those of the penetrating one at equal angles with both
the planes of projection. Then to find their lines of penetration, draw
in first the plan and elevation of two pyramids—having equal bases
and altitudes—with their axes at right angles. The combined triangle
and square shown in No. 1, and the three triangles shown in No. 2
(Fig. 176) will be the plan and elevation of the solids—as if they were
entire—in the given positions.

Now an *axial* section of both solids by a vertical plane along the
line $a\,b$, No. 1, will give a triangle for each, and points 1, 2, 3, 4, No.
2, as the extreme ones in the penetration. Any other vertical section
of the solids, taken through the apex b of the penetrating one, will
also give a triangle for the section of each. If then a plane, coinciding
with the *front* face fb of the penetrating pyramid, be passed through
the penetrated one, it will give in No. 2 the triangle $c\,d\,e$—partly
shown in faint lines—as the section; the parts of its two sides—
between $d'c'$ and $d'e'$, shown in full—being the lines of penetration
seen in elevation. To find those seen in plan, let fall from points 1, 3,
d', d'' in No. 2, projectors to cut the corresponding front and back
edges of the pyramids in 1, d', d'', 3, No. 1; join these points by
straight lines as shown, and they will be the lines of penetration of
the top face of the horizontal pyramid with the slant sides of the vertical one. To find the similar lines on the *under* face of the same
pyramid, project over from No. 2, to the plan No. 1, the points 2, c', e',
4; join these by dotted lines, as shown, and they will be the ones
required.

Assuming the axial plane $a\,b$ of the two pyramids to be at an
angle, say of 30° with the VP, but still perpendicular to the HP, and
an elevation of the solids in that position to be required; then the
projection shown in No. 4 (Fig. 176) will be what would be obtained,
as the vertical pyramid has been moved round on its axis until the
axial plane of both solids makes the required angle with the VP.
The elevation and lines of penetration have then been found by direct
projection from the plan No. 3 and elevation No. 2.

The case in which two equal square pyramids penetrate each other
in such a way that both have one of their greatest plane sections in one
and the same plane, their axes being at right angles, is shown in Nos.
5 and 6 (Fig. 176).

69. Now, the greatest plane section of a square pyramid is evidently
one obtained by cutting it through from apex to base in such a way as
to divide its base *diagonally* into two equal triangles. Then, if two
such pyramids as A and B (Fig. 176), equal in size, and at right angles
to each other, be cut simultaneously by a plane passing through their
axes and cutting their bases diagonally, their elevation, assuming them
to be entire, and both axes parallel to the VP, will be as shown in No.
5, but without the lines of penetration 1, 2, 3; 4, 5, 6. To find these
lines, first draw in the plan (No. 6) of the pyramids as if entire.
Assume a vertical plane, tangent to the front edge $x\,b$ of pyramid B, to
cut through A to its base. It would intersect that base in its two front
edges at points f, g, and its front slant edge in point i. Obtain by
projection the elevation of the triangular section f, i, g thus produced—

160   FIRST PRINCIPLES OF

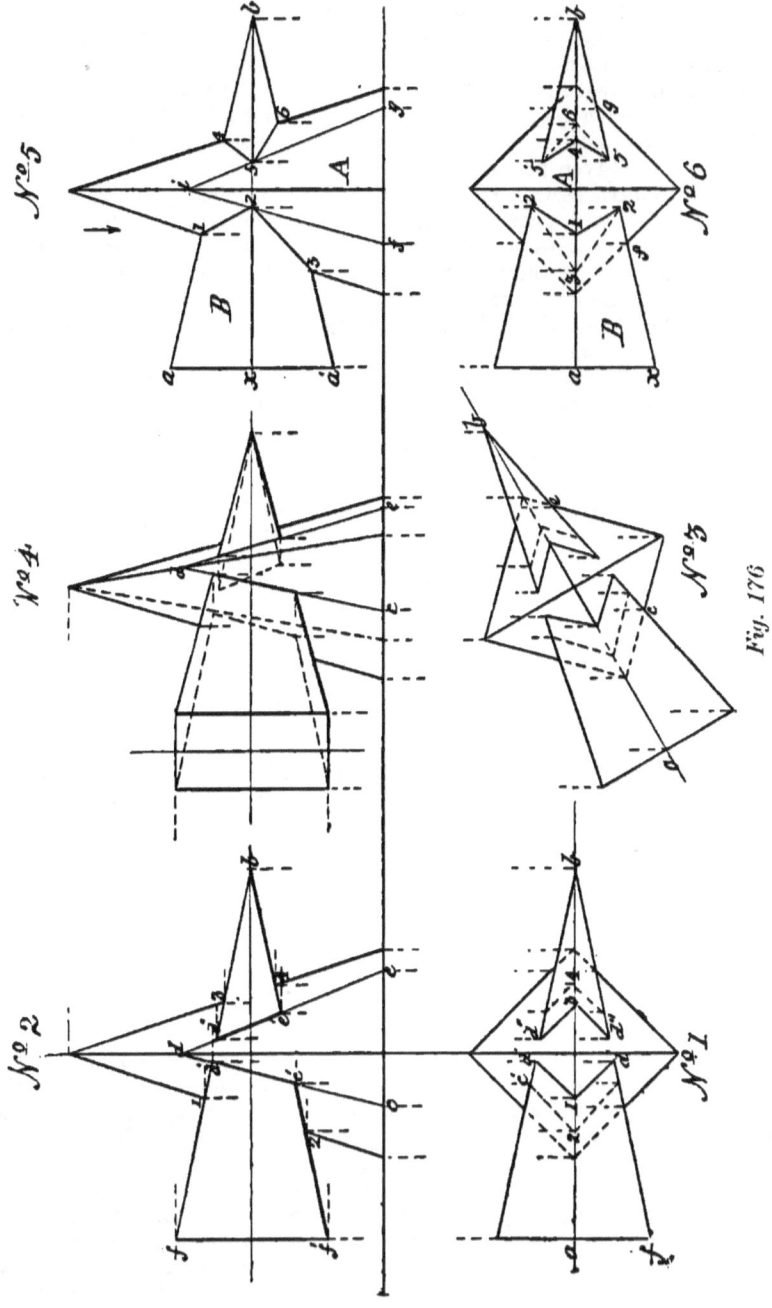

Fig. 176

MECHANICAL AND ENGINEERING DRAWING 161

shown in faint lines in No. 5—and it will be found to cut the front edge xb of the prism B in points 2, 5; join these by straight lines to points 1, 3 and 4, 6—previously found—and they will be the required lines of penetration of the two solids seen in elevation.

To find the plans of these lines, let fall projectors from points 2, 5 and 1, 4, in No. 5, to cut the corresponding edges of the prism B in No. 6, and they will give points 1, 2, 4, 5; join these as before by straight lines as shown, and they will be the plans of the two lines of penetration 1, 2, 4, 5 previously found in No. 5.

On looking upon the two pyramids, in the direction of the arrow shown in No. 5, two *pairs* of lines of penetration will be seen. These are the two front ones 1, 2; 4, 5; and the two, 1, 2'; 4, 5', immediately behind them, shown in No. 6. To show the position of the four lines of penetration made by the two *lower* inclined sides of pyramid B, let fall projectors from points 3, 6 in No. 5, to cut the corresponding edge of the same pyramid in 3, 6 in No. 6; join these points by dotted lines to 2, 2'; 5, 5' respectively, and they will be the plans of the lower, or *return* lines of penetration of the two solids.

Should the two prisms have such a relative position that their axes are inclined to each other at some angle other than a right angle, in a plane common to both, the procedure for finding their lines of penetration would still be the same. As a test of its application, let the two pyramids in the last problem penetrate each other in such a way that the axis of the penetrating one is inclined at an angle of 30° to the horizontal, while that of the penetrated remains vertical, and let it be required to find their projection and lines of penetration when the axial plane of the solids is parallel to the VP, and also when that plane is inclined to the VP at a given angle.

As the solids in the first-named position have their axial plane parallel to the VP, and the axis of one of them is inclined to the HP, their elevation, as if they were entire, must first be drawn. Having done this as shown in No. 1 (Fig. 177), from it find by projection their plan—also as if entire—as given in No. 2. The points of intersection of the upper and lower edges of the pyramid B with A are at once seen in No. 1, to be 1, 2, 3, 4. Proceed as in the previous problem to find the section of A, by a vertical plane passing through it tangent to the front edge of B, and the points in that edge cut by it; join these up to the first found points by straight lines as before, and they will be the lines of penetration in elevation. On finding the plans of these lines as before explained and shown in No. 2, the first part of the problem is solved. For its latter part, draw in as shown in No. 3 in the figure, a repeat of No. 2, but having its axial line ab at the required angle—say 30°—with the VP or IL; and from it and No. 1 find by projection the elevation of the pyramids and lines of penetration in their new position. This view, if obtained in accordance with the procedure already explained in a previous problem, will give No. 4 (Fig. 177) as the result.

70. As a concluding problem in the penetration of plane-surfaced solids, we give in No. 5 and No. 6 (Fig. 177) a case in connection with pyramids which sometimes occurs in practice—viz., that wherein a

M

162   FIRST PRINCIPLES OF

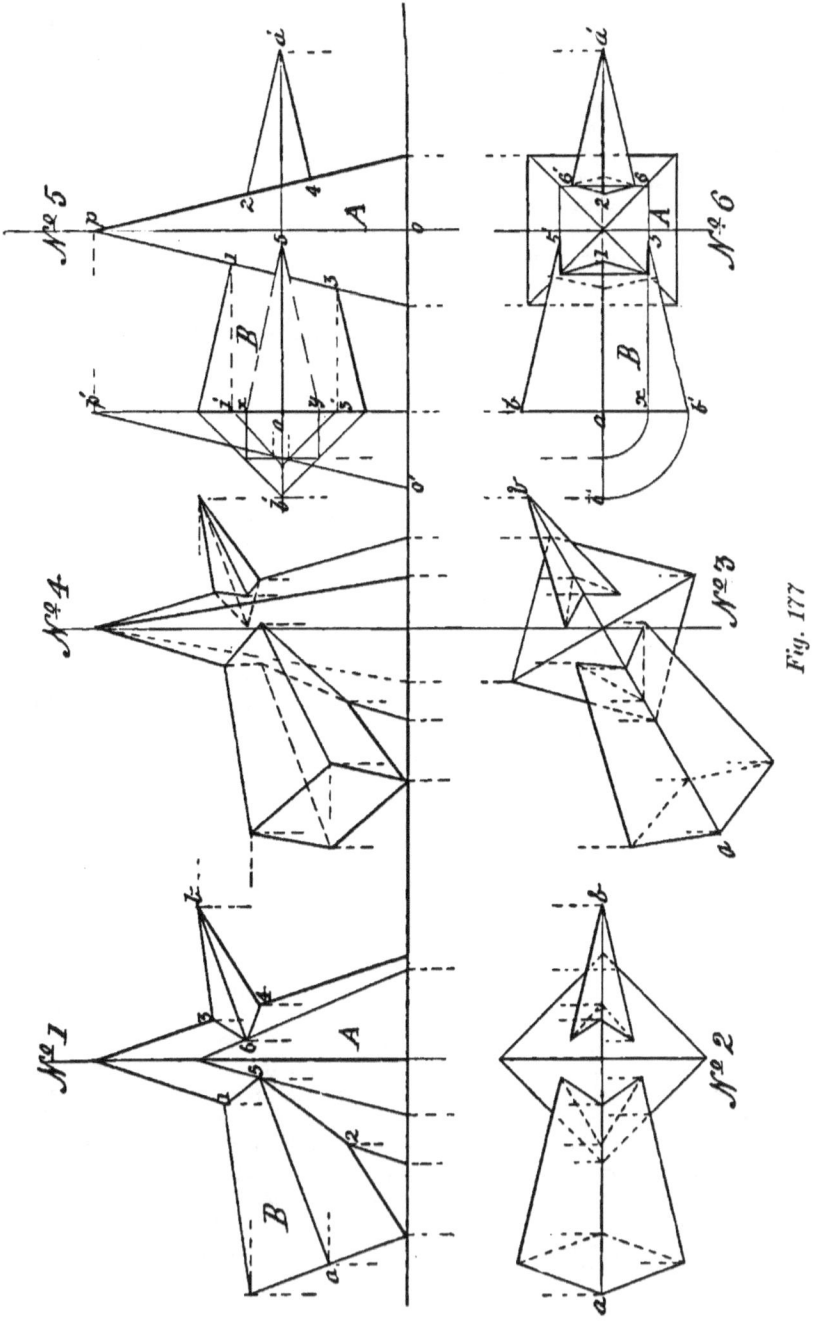

Fig. 177

MECHANICAL AND ENGINEERING DRAWING 163

square pyramid is penetrated by a similar solid, the relative position of the two being such that an axial plane passed through *both* gives a similar section in each, but divides them into dissimilar solids. The problem is—

Problem 74 (Fig. 177).—*A square pyramid A, with its axis vertical and base edges parallel and at right angles to the VP, is penetrated by a similar pyramid B, having its axis horizontal and parallel to the VP; required the projections—plan and elevation—of the lines of penetration of the two solids, when the diagonals of the base of the penetrating one are respectively parallel and perpendicular to the VP.*

First draw in the plan and elevation of the pyramids—as if they were entire—in the position stated in the problem, and shown in Nos. 5 and 6 in the figure. The extreme points of intersection of the top and bottom *edges* of the pyramid B, with the right and left faces of A, are at once seen to be 1, 2, 3, 4, No. 5. Then to determine where the front and back edges of B enter and leave A, assume a *horizontal* plane to pass through *both* solids on the line of the axis—$a\,a'$—of B, and the parts cut off by it to be removed. The sections exposed will be a triangle $b\,a'b'$, for B, and a square—in faint lines—for A, shown in No. 6.

Now the two sides, $b\,a'$ and $b'a'$, of this triangle are the *plans* of the front and back edges of the pyramid B, and as the square is a section of the pyramid A made by the same cutting plane, it is at once seen that the two horizontal edges of B enter the pyramid A at points 5, 5′ in its *front* and *back* faces, and leave it at points 6, 6′ in its right face. Project over point 5 in $b'a'$, No. 6, to 5 in $a\,a'$, No. 5. Then to find the direction of the lines of penetration—which will meet in point 5 in No. 5—get a side elevation of one-half of the pyramid A, using the base line of B produced to p' as an axis. On this line find by projection a half-section of the pyramid B, made by a plane passing through points 1 and 3 on the left face of the pyramid A, which will be a triangle, having the line 1′3′ for its base, and its two sides cutting the slant side $p'o'$ of the half-pyramid previously found in two points. These projected over to the left edge of the pyramid A, and joined to point 5 in $a\,a'$ as shown, are the lines of penetration sought. As the finding of the plans of these lines, as shown in No. 6, require only the careful application of principles already explained, they are left to the student to draw in without further assistance.

As the lines of penetration of prisms and pyramids by similar solids, *having any number of sides*, are found by a similar procedure to that so fully explained in this and the preceding chapter, the subject of the penetration of solids having *curved* surfaces—a most important one to the engineering draughtsman—will next be considered.

CHAPTER XIX

THE INTERSECTIONS OF SOLIDS HAVING CURVED SURFACES

71. As the cost of any engine, or machine detail, depends in a great measure upon the number of processes involved in its production, it will readily occur to the student of engineering that the simplest possible form, and one requiring the least amount of machining, to fit it for the purpose to which it is to be applied, will be that which is produced by a *revolving* motion of the material to be acted upon by the cutting tool which gives it shape. The product of this motion, a "solid of revolution," has also a larger cross-sectional area within the same bounding line than any plane-surfaced solid; and for this reason, it will be found on studying the construction of engines, machines, and all ordinary steam generators, that the cylinder, the cone, and the sphere—all of them solids having curved surfaces—enter much more largely into their design than do any other of the solids. A knowledge of the way in which their surfaces intersect when in combination is therefore highly essential, not only to the draughtsman who is called upon to design such details, but to the skilled workman who will produce them.

This knowledge is the more necessary, as upon the correct application of the principles of projection to the determining of the *actual* intersections of such solids, depends the truth or otherwise of the results arrived at when finding the "developments" of their surfaces—a subject which immediately follows that now being considered, and one which includes within its scope the determination of the exact shape the material should take when the article to be produced is one made entirely of metal plates.

The preliminary problems in connection with the intersections of curved-surfaced solids are those in which cylinders are penetrated by cylinders. The first is—

Problem 75 (Fig. 178).—*Given the plan of two equal-sized right cylinders, with axes at right angles in one plane parallel to the VP, and penetrating each other; to find their lines of intersection.*

Now, from the way in which a right cylinder is generated, it is known that an *axial* section of the solid is a rectangle, and any other

164

MECHANICAL AND ENGINEERING DRAWING   165

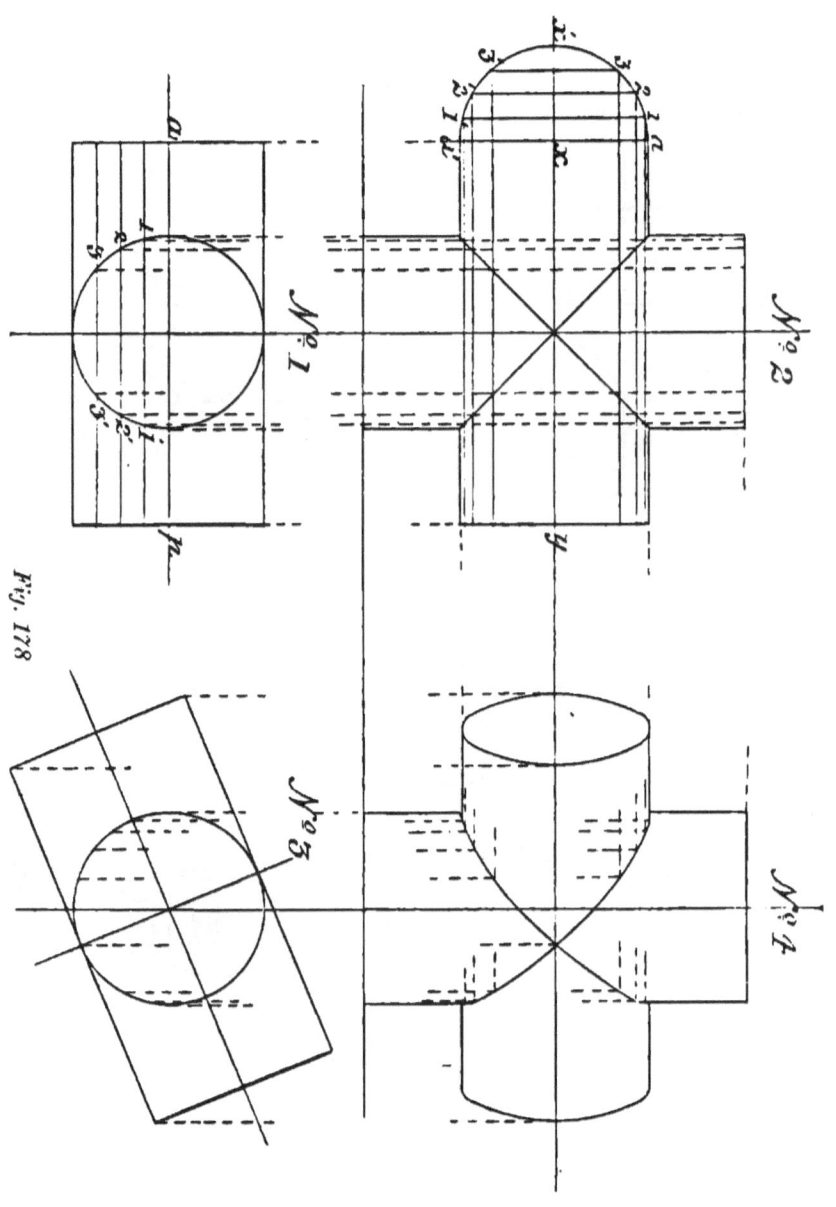

Fig. 178

166   FIRST PRINCIPLES OF

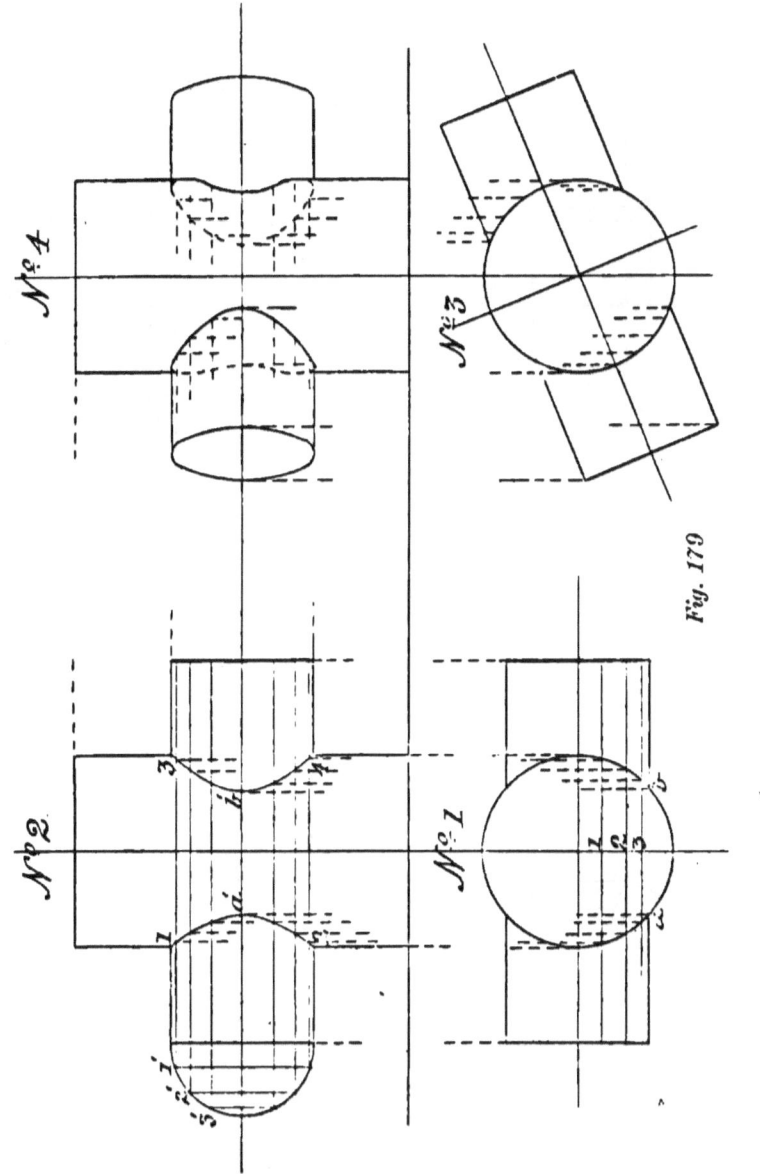

Fig. 179

MECHANICAL AND ENGINEERING DRAWING

section parallel to an axial one is a similar figure. Therefore, if No. 1 (Fig. 178) be the given plan of the two solids, to find the lines of intersection of their surfaces, first get an elevation of them as if entire. As only the *front* halves of the two cylinders will be seen in elevation, it is with them that we have to deal. If, then, a series of vertical section planes parallel to each other and the axial plane *a p* of the solids, be assumed to pass simultaneously through their front halves, the points of intersection of the sections of the cylinders by these planes will be points in their lines of penetration.

To find these lines, draw in No. 1 (Fig. 178) parallel to the line *a p*, as many lines—say three—at convenient distances apart, as it is intended to use section planes. Find by projection from the points 1 1′, 2 2′, 3 3′—where these lines cut the *plan* of the vertical cylinder—the sectional elevations produced by them, as shown in dotted lines in No. 2. For the corresponding sectional elevations of the horizontal cylinder by the same planes, on its axial line *x y*, No. 2—produced to the left—with point *x* as centre and *x a* as radius, describe the semi-circle *a x′a′*. This will be one-half of the end of the horizontal cylinder turned on its vertical diameter *a a′* as a hinge. In this semi-circle set off from *x*—in the line *x x′*—the distances that the lines 1 1′, 2 2′, 3 3′ in No. 1 are from the line *a p*, and through the points thus found draw lines parallel to *a a′* to cut the semi-circle in points 1 1′, 2 2′, 3 3′. Faint lines drawn through these points from end to end of the horizontal cylinder, parallel to *x y*, or the IL, will give the corresponding sections of it made by the same planes as used in the vertical cylinder. Then the points where the edges of the corresponding sections of both cylinders cut each other, as shown in No. 2, are points in their lines of intersection, which, when joined, will be found to be straight ones crossing each other at right angles.

The student will note that in this case the cylinders being *equal* in diameter and at right angles to each other, their intersections form true mitres dividing them into four equal parts, each having semi-elliptic sectional surfaces perpendicular to each other.

In the case of two equal cylinders intersecting at right angles, and having the plane of their axes at an angle to the VP, but still in the vertical position, their lines of intersection would be found in the same way as shown in No. 2; the only difference in the result being that such lines will be curved instead of straight, showing, as will be seen in No. 4 (Fig. 178), that they become, through the altered position of the solids with respect to the plane of projection, portions of ellipses. As the finding of the elevation and lines of penetration of the two cylinders, in their new position, offers to the student some practice in the projection of ellipses, which will very frequently occur in this part of the subject, no further assistance is given in this problem than that afforded by the few projectors shown in the diagram.

72. The next case is one where the intersecting cylinders are *not* of equal diameter, but still have an axial plane common to both. The problem is—

168   FIRST PRINCIPLES OF

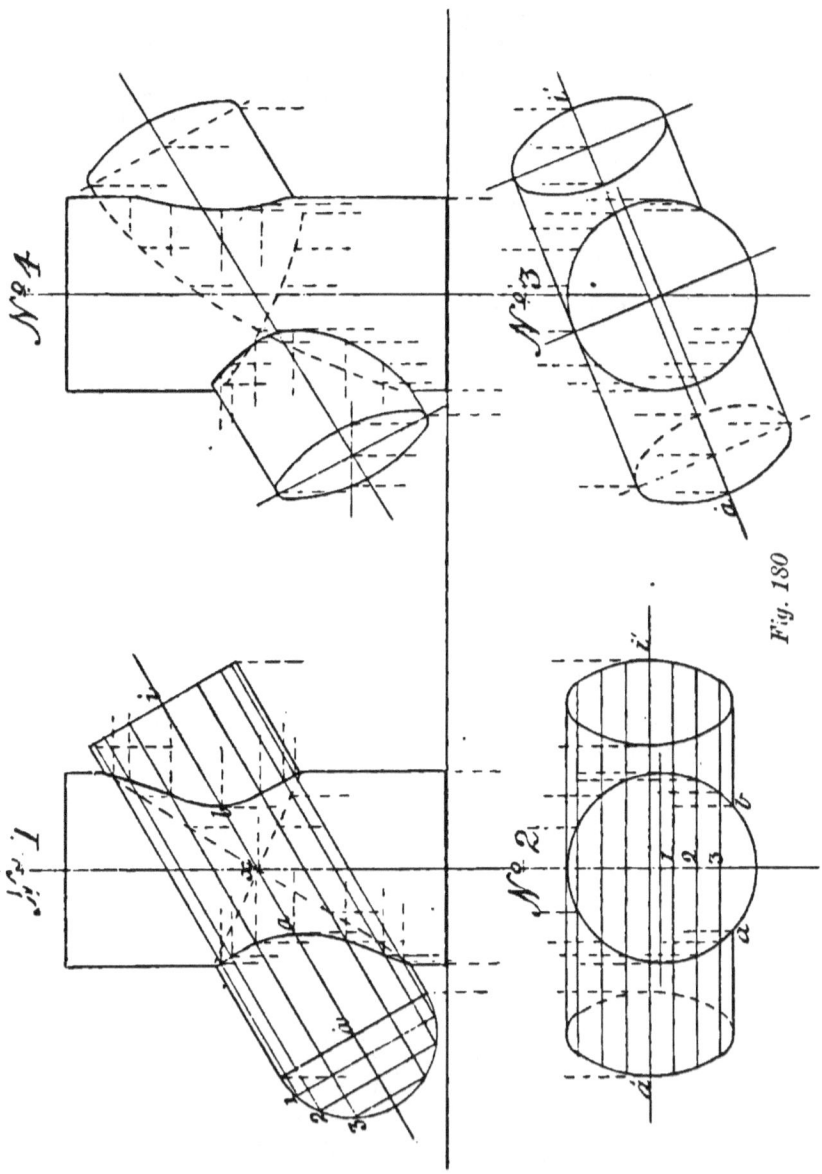

Fig. 180

MECHANICAL AND ENGINEERING DRAWING   169

Problem 76 (Fig. 179).—*Given the plan of two cylinders of unequal diameter intersecting each other, with their axes at right angles and in one plane parallel to the VP ; to find their elevation and lines of penetration.*

The plan of the two cylinders in the position stated in the problem will be that shown in No. 1 (Fig. 179), the penetrating one being horizontal. From this plan, find by projection the elevation of the cylinders as if entire, making them both of the same length, with the horizontal intersecting the vertical one at the middle of its height. Then to find the visible lines of intersection of the solids in elevation, draw in—as in the last problem—the plans of the intended vertical section planes to be used ; and obtain by projection the elevation of the sections of the vertical cylinder made by these planes. Also draw in, in No. 2, on the axis of the horizontal cylinder produced, an end view of the front half of that cylinder. On this, show in—as in the previous problem—the lines made by the vertical section planes in passing through that cylinder, and from the points in the semi-circle, or end view, cut by these lines, obtain as before the front elevation of the sections of the horizontal cylinder ; then the points where these sections cut the corresponding ones of the vertical cylinder, will be points in the lines of penetration of one solid by the other. On joining these by lines drawn through them, they will be found to be (as shown in No. 2, Fig. 179) symmetrical curves ; the highest and lowest points in them being determined by the intersections of the bounding lines of the two solids, at 1, 2, 3, 4, while the middle ones, $a'b'$, are the vertical projections of $a\,b$ in No. 1, or where the frontmost part of the horizontal cylinder is seen to enter and leave the vertical one.

Should the axial plane of the two cylinders be at any given angle with the VP—as in No. 3—instead of being parallel thereto, the lines of their intersection would still be found in the same way as before shown, but they will not be symmetrical curves as in the problem just solved, but such as are seen in *full* lines in No. 4 (Fig. 179), the dotted ones being the return lines of intersection of the penetrating —or horizontal—cylinder, with the surface of the vertical one, which is out of sight. As the correct finding of the intersecting lines—shown in full, and dotted in No. 4—of the two cylinders is but a matter of careful projection by the student, no further assistance than that given in the diagram will be required by him.

73. The next case is that of the intersections of unequal cylinders, at an angle other than a right angle, and having their axes in different planes at a given distance apart. The problem is—

Problem 77 (Fig. 180).—*Two cylinders of unequal diameter, with their axes in parallel planes, intersect each other at an angle of 30° ; it is required to find the projections and lines of penetration of the solids, when the axis of one of them is vertical, and the other parallel to the VP.*

170   FIRST PRINCIPLES OF

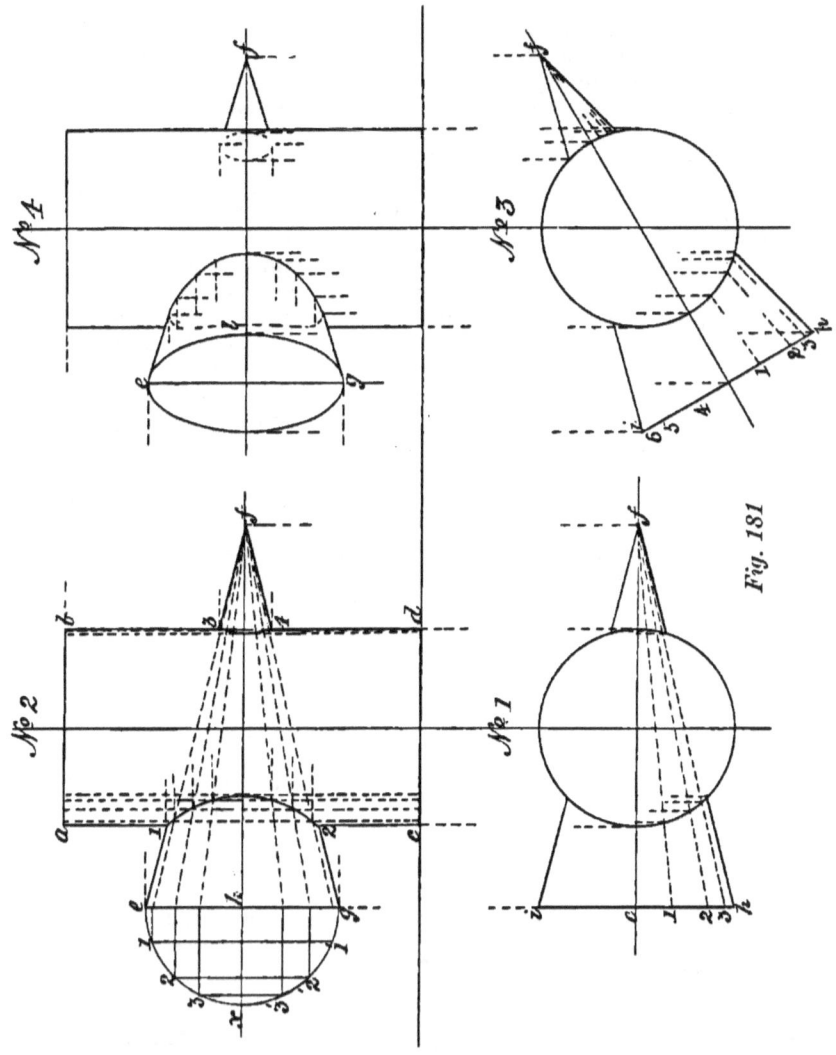

Fig. 181

MECHANICAL AND ENGINEERING DRAWING   171

As one of the cylinders is inclined, an elevation of both as if they were entire must first be drawn, as shown in No. 1. From this obtain by direct projection the plan of the solids as given in No. 2. Then to find the visible lines of intersection, assume parallel planes represented by the lines 1, 2, 3, in No. 2, to pass vertically through both solids. An elevation of their sections made by these planes will show where they intersect, and will give the points through which the required lines will pass. For the *return* lines of intersection, or those out of sight—shown dotted in No. 1—the same method of procedure is repeated as in finding those which are seen. As a vertical plane would be tangent to the extreme *back surface* of both cylinders, the point x in the elevation No. 1 would be where the lines of penetration of the two solids would cross. A plane passing through the vertical cylinder tangent to the *front* surface of the inclined one gives the two points $a\,b$ in plan and elevation, as those where that surface enters and leaves the vertical cylinder. Should the plane of the axis of the inclined cylinder, represented by the line $a\,i$ in No. 2, be otherwise than parallel with the VP —say, as in No. 3—the lines of intersection of the two solids would be found in the same way as that explained in Fig. 179, the difference in their appearance—as seen in No. 4—being the result of the changed position of the solids with respect to the plane of their projection, the vertical cylinder having been turned on its axis through a certain angle, carrying the inclined cylinder with it. With the assistance of the few projectors shown, and bearing in mind that *each* point in the lines of intersection found in No. 1 has, by the turning of the vertical cylinder, passed through the *same* angle horizontally, the student should be able to find without trouble the actual lines due to the altered position of the solids which are shown in No. 4 (Fig. 180).

As the intersections of cylinders in *any* position only require the correct use of section planes as exhibited in this chapter to determine them, no further examples are given in this connection, but we pass on to the solution of one or two problems in which the intersections of the cone and cylinder are involved.

Problem 78 (Fig. 181).—*A cylinder in a vertical position is penetrated at the middle of its height by a cone, having its axis horizontal and parallel to the VP ; to find the lines of intersection of the two solids when their axes are in one plane.*

Let No. 1 (Fig. 181) be the given plan, and No. 2 the elevation of the two solids ; then from their relative position it is evident that an axial plane passed through both of them will give the rectangle $a\,b\,c\,d$, and the triangle $e\,f\,g$, as their respective sections. The extreme points in the penetration of the cylinder by the cone will be 1, 2, 3, 4. Now, although a series of vertical planes passed through both solids, parallel to this axial one, would give rectangular sections of the cylinder, those of the cone by the same planes, instead of being triangular, would be hyperbolic, necessitating a great amount of careful projection in finding them. To obviate this unnecessary trouble the same results may be arrived at by a much more simple procedure. Instead of assuming a

172   FIRST PRINCIPLES OF

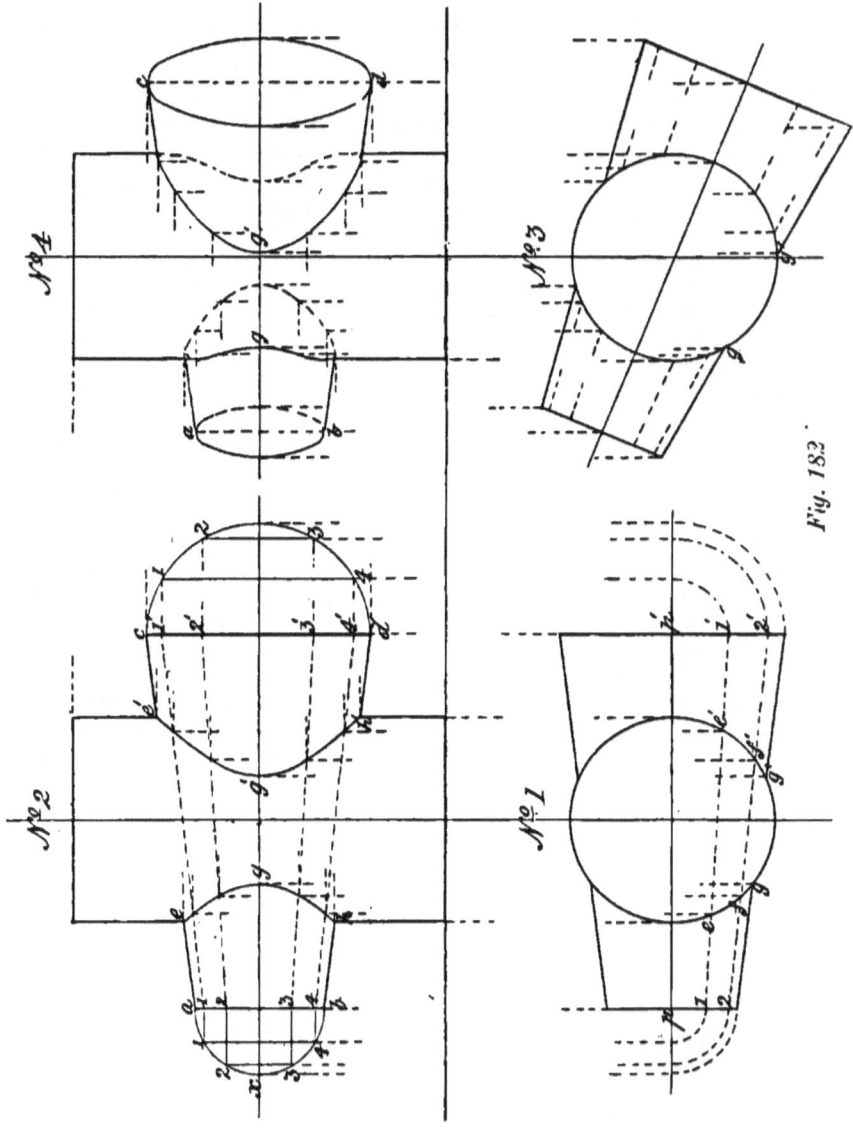

Fig. 182

MECHANICAL AND ENGINEERING DRAWING   173

number of *parallel* planes to pass simultaneously through both solids, all
the points of the intersection required may be found by making the
cutting planes *divergent*.

74. In a previous chapter it was shown that all the sections of a cone
made by a plane passed through its apex to its base, gave triangles for
those sections. If, then, in the problem under solution, cutting planes
are assumed to pass through the apex of the cone in varying directions,
cutting it and the cylinder simultaneously, it is evident that *all* the
sections of the *two* solids will be similar to those obtained by the axial
section—viz., rectangles and triangles, each of which will give four
points in the lines of intersection sought.

To find these points in No. 2 (Fig. 181), through *f*, the apex of the
cone in No. 1, draw the lines *f*1, *f*2, *f*3, and let them, with *e f* and *h f*,
be the plans of vertical section planes passing through both solids in the
directions shown. Then find by projection the rectangular sections of
the cylinder produced by the cutting planes, as shown in faint dotted
lines in No. 2. For the corresponding sections of the cone by the same
planes, on its base line *e g* in No. 2 as a diameter, and *h* as centre,
describe the semi-circle *e x g*, which will be the front half of the base of
the cone looked at from the left. On the line *x h*, measuring from *h*,
transfer the distances *e* 1, *e* 2, *e* 3 in No. 1, and through the points thus
found draw lines parallel to *e g*, to cut the semi-circle in 1 1', 2 2', 3 3';
project these points over to *e g*, and through them draw faint lines to
the apex *f*. The triangles formed by these lines will then be the
sections of the front half of the cone, made by the same planes that cut
through the cylinder, and the points of intersection of these respectively
will be the points in the lines of penetration sought.

Should the axis of the cone be inclined to the V P, as in No. 3, that
of the cylinder still remaining vertical, and the lines of intersection be
required, the process of finding them is exactly the same as before. The
lines will, however, assume a different form of curve, due to the altered
position of the cone with respect to the plane of its projection, the V P.
This alteration, it will be noted, brings the base of the cone into view,
but causes the surface of the cylinder penetrated by its *small* end to
pass out of sight. As the axis of the cone in the previous figure was
parallel to the VP, the lines of intersection of its back half with
the surface of the cylinder being directly behind those in front, could
not be shown; but the cone being now inclined to the VP, it is possible
to indicate exactly the form of the return curve of penetration. This is
shown by the dotted line *l*, obtained by projecting over the points 4, 5,
6, in *i h*, No. 3, and finding as before the corresponding sections of the
back halves of the cone and cylinder, which give the points required.

75. Had the penetrating solid been but a *portion* of a cone—say a
frustum—then the lines of intersection of the two may be got in two
ways. If the apex of the cone is *accessible*, the method already explained
would be the simplest and best; but if the *taper* of the cone be only
slight, and its apex *out of reach* or at an inconvenient distance, then the
procedure would be as shown in Fig. 182.

First, let it be assumed that the axis of the penetrating solid is
parallel to the VP, and at right angles to that of the cylinder. Having

174 FIRST PRINCIPLES OF

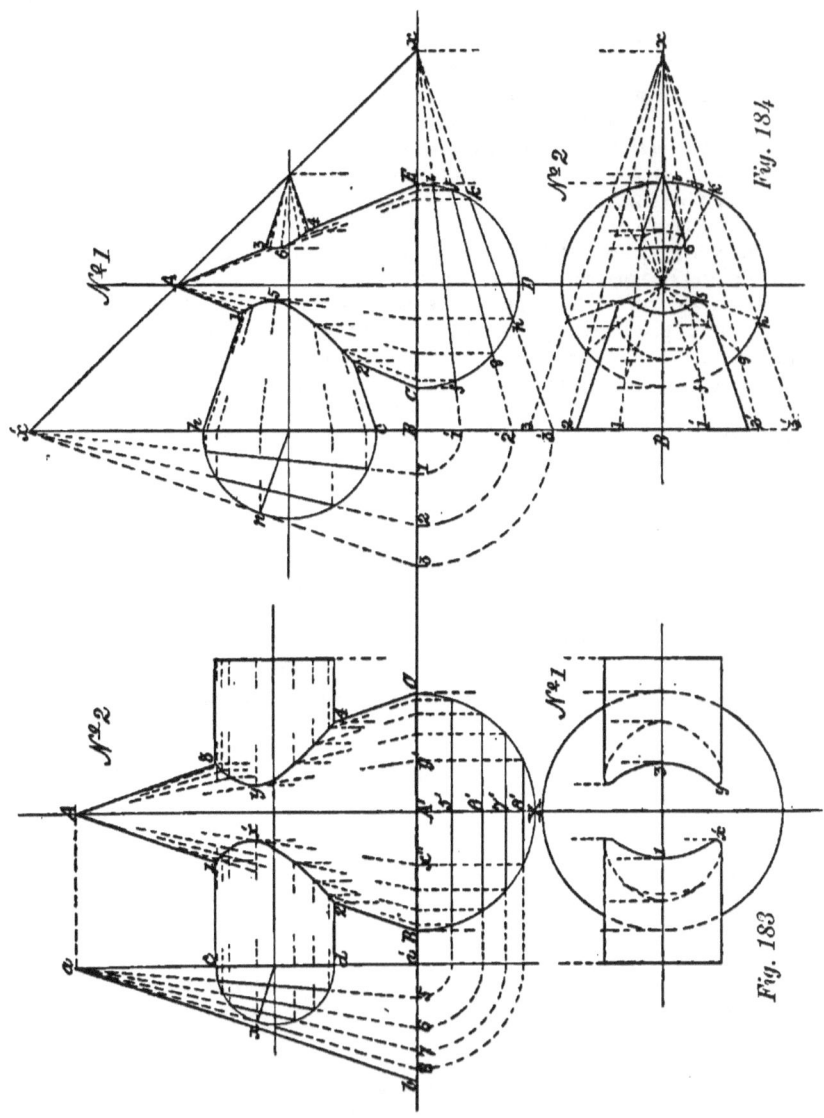

Fig. 184

Fig. 183

MECHANICAL AND ENGINEERING DRAWING   175

drawn in the plan and elevation of the solids as if entire, to find
their lines of intersection divergent planes are assumed to cut through
both, as in the last case; but to ensure that the planes used will pass
through the *apex* of the cone—of which the frustum forms part—and
thus give true sections, the procedure in finding them is as follows :—

In No. 2 (Fig. 182), on $a\,b$, $c\,d$ (the ends of the frustum), as diameters,
describe semi-circles ; divide each into the same number of equal parts—
say, six—and project the points of division in them over to their
respective diameters ($a\,b$, $c\,d$), and join them by faint dotted lines, as
shown. Then the surfaces included within the bounding lines 1 1', 1' 4',
4' 4, 4 1, and 2 2', 2' 3', 3' 3, 3 2, will be the true sections of the front half
of the frustum, made by the two section planes passing vertically
through it, at distances from its axis at either end equal to that which
the vertical lines drawn through points 1, 4, and 2, 3 in *each* semi-circle
are from the lines $a\,b$ and $c\,d$ respectively.

To find the sections of the cylinder by the same cutting planes, set
off at the corresponding ends of the frustum in the plan No. 1 the
distances that these planes are from the axial one $p\,p'$ of the two solids,
and through the points 1 1', 2 2', thus found, draw faint lines cutting
the plan of the cylinder at $e\,e'$, $f\,f'$. Obtain by projection the elevation
of the sections of the cylinder at these lines, and the points where they
cross the corresponding sections of the frustum—in elevation—will be
points in the lines of intersection sought. To find the two points in the
cylinder's front surface, where the frustum penetrates and leaves it,
draw projectors through $g\,g'$ in the plan No. 1, and they will give $g\,g'$
in the elevation ; through these and the points of intersection of the
sections—already found—draw the curved lines $e\,g\,h$, $e'\,g'\,h'$, and they
will be the lines of penetration of the cylinder by the frustum of the
cone.

If the axis of the penetrating solid is *inclined* to the VP or HP, or
to both, that of the cylinder still remaining as before, as shown in No. 3
(Fig. 182), the lines of intersection of the two are found as in No. 4
(Fig. 181)—viz., by direct projection from the plan—No. 3—of the
solids in their new position, and from the elevation No. 2 ; the result
being the view given in No. 4. The return lines, or those on the surface
of the cylinder *nearest* the VP shown dotted, are found as indicated by
the projectors.

Taking the converse position of the two solids—which frequently
occurs in practice—or that where a cone is penetrated by a cylinder,
for a further problem in this connection, it would be solved on the same
principles, though in a different way.

Problem 79 (Fig. 183).—*A cone with its axis in a vertical position
is penetrated by a cylinder ; required the lines of intersection of the
solids in plan and elevation, when the axis of the cylinder is horizontal, and parallel to the VP, and in the same plane as that of the
cone.*

Draw in first the plan and elevation of the solids as if entire. Then
as the axes of both are in one and the same plane, an axial section

176   FIRST PRINCIPLES OF

of them will at once give 1, 2, 3, 4 in No. 2 as the extreme points in
the penetration. To find others through which the lines sought will
pass; at one end of the cylinder—say the left—give a side elevation of
the front halves of the two solids as shown. Assume the right-angled
triangle $a\,a'\,b$, and the semi-circle on $c\,d$, to be half-sections of both
solids swung round on the axis of the cone until parallel with the VP.
From these it will be seen that a part of the front and back surfaces
of the penetrating cylinder lie wholly within the cone. To find how
much, and thus determine the lines of intersection sought, proceed as
follows :—

Through the vertex of the right-angled triangle $a\,a'\,b$, draw straight
lines to its base ; the first one tangential to the semi-circle on $c\,d$, and the
others cutting it at suitable points, as shown. These lines are the edge
views of the section planes passing through both solids, which will give
the points required, as the sections produced by them will be triangles
for the cone and rectangles for the cylinder. The rectangles are at once
found by drawing lines through the points in the semi-circle (cut by the
assumed section planes) parallel to the axis of the cylinder.

For the corresponding triangular sections of the cone, describe on its
base B C as a diameter the semi-circle B X C. On its axis, A A'
produced, set off from A' the several distances that the points 5, 6, 7, 8
in the base of the triangle $a\,a'\,b$ are from the point a' in it. Through
the points 5', 6', 7', 8, thus found, draw lines parallel to B A' C, and from
where they cut the semi-circle find by projection the elevation of the
triangular sections of the cone, as shown in No. 2 ; then the points
where the corresponding sections of the two solids intersect are points
in their lines of penetration. The two points, $x'\,y$ in those lines—or
where the plane drawn through $a\,8$ in the side elevation is tangent
to the surface of the cylinder—are found by drawing a line through x,
where the plane $a\,8$ touches, and intersecting it with the triangular
section A $x''\,y'$ of the cone made by the same plane.

The *plans* of the lines of intersection $1\,x\,2$ and $3\,y\,4$ in No. 2
are most easily determined by finding horizontal sections of the cone—
which will all be circles—at several points in those lines in elevation,
and letting fall projectors from them on to the circular sections so
found ; then the points where the projectors cut these sections will be
those through which the required lines of penetration are to be drawn.
As the solids are directly at right angles to each other, their lines of
intersection are consequently symmetrical on either side of the cone,
both in plan and elevation.

76. As the procedure in finding the lines of intersection of a cone
and cylinder having their axes in the same plane, but *not* at right
angles, would be the same as in the last example, as any plane passing
through the apex of the cone to its base, and simultaneously cutting
through a cylinder parallel to its axis, however inclined, would still give
a triangle as the section of the former, and a rectangle as that of the
latter, and the points of intersection of the sectional surfaces so
produced as points in their lines of penetration, we pass on to the next
combination of solids employed in giving form to engine and boiler
details—viz., that of cones with cones, a knowledge of their intersections

MECHANICAL AND ENGINEERING DRAWING

being the more important, as combined conical surfaces enter into almost all designs where metal plates are used as the material of construction. As a problem in this connection take—

Problem 80 (Fig. 184).—*A cone with its axis in a vertical position is penetrated by another cone having its axis horizontal and parallel to the plane of projection; required the lines of penetration in plan and elevation of the two solids in the position given.*

Draw in first the plan and elevation of the two cones as if entire. Then, as a series of planes passing through the apex of one of the cones to its base would give triangles for the sections made of that particular cone only, but hyperbolas for those of the cone in combination, it is evident that the finding of the lines of intersection of the two solids by the use of section planes in such a way would be a long process and very liable to error. What has to be aimed at, is to so use a section plane that it will, in the problem to be here solved, give *similar* sections of both cones. To effect this, a *triangular* plane is assumed to pass through the apices of both simultaneously, and to be swung on one of its sides—its hypothenuse—as a hinge through varying angles.

To apply this principle to the problem to be solved (in the elevation of No. 1, Fig. 184), produce the base line hc of the horizontal cone indefinitely in both directions; then draw a line through the apices of *both* cones to cut the produced base line in the point x' and the IL in point x. Then will the produced base of the horizontal cone from v' to the IL at B be the *perpendicular*, the line Bx the *base*, and that drawn through the apices of the cones from x' to x the *hypothenuse*, of the right-angled triangular section plane for finding the lines of intersection of the two solids. Such a plane in its normal position —or parallel to the VP and perpendicular to the HP—at once gives 1, 2, 3, 4 in No. 1 as the extreme points in the lines of penetration. To find others through which these lines will pass, the positions of the triangular plane $x'Bx$ in its swinging on the line $x'x$ as a hinge have next to be shown.

To do this, describe on the base hc of the horizontal cone as a diameter, a semi-circle, then through the vertex x' of the right-angled triangle $x'Bx$, draw straight lines, the outermost one $x'3$ tangential to the semi-circle, and the others, $x'2$, $x'1$, at about equal distances apart. These lines are the *end* elevations of the perpendicular edge of the triangular plane $x'Bx$, when swung on its hypothenuse $x'x$ as a hinge. Its *outer* position, represented by the line $x3$, is drawn tangent to the front surface of the horizontal cone, to find the points 5 and 6 in the lines of penetration. To determine their exact position and that of other points in these lines, describe on the base CE of the vertical cone as a diameter, the semi-circle CDE; transfer the points 1, 2, 3 on the IL by arcs, as shown, to 1', 2', 3', on the produced base line of the horizontal cone, and through these last-found points draw lines to x, intersecting the semi-circle last drawn in the points f, g, h, and i, j, k. Find by projection from these points and those made by

N

FIRST PRINCIPLES OF

the same section plane, in the semi-circle drawn on the base hc of the horizontal cone, the corresponding vertical sections of the two cones, and the points where they intersect will be points through which their lines of penetration will pass, as shown in No. 1 (Fig. 184). To find the plans of these lines, the simplest method, and one involving the least confusion of lines in the diagram, is that shown in No. 2, where the triangular sections of the vertical cone made by the planes $x'3$, $x'2$, $x'1$, are first found, the points required being then projected over from No. 1 to their corresponding sections in the plan, and lines of penetration drawn through them, as shown in No. 2.

Should the axis of the penetrating cone be at an angle with the VP, and an elevation of the solids with their lines of penetration be required, this would be obtained by transferring the plan of them already found in No. 2, to such a position in the HP as would bring the axis of the horizontal cone to make the required angle with the VP or IL; the required elevation is then found by direct projection from the plan of the solids No. 2 (in their new position) and the elevation given in No. 1.

In finding such a view of the two solids, the student is to be reminded that great care is necessary in projecting over the various points in both plan and elevation, so as to ensure the correct representation of the solids themselves, as well as their lines of intersection.

77. As the same procedure would be followed in finding the lines of penetration of two cones, when the axis of the penetrating one is *inclined* to both the VP and HP—the only difference being, that the triangular section plane used in finding the points in the lines of intersection of the solids in the first part of the solution, would not be right-, but acute-angled—the student is left to work out such a problem himself, and we now pass to the finding of the lines of penetration or intersection of the sphere, with such of the solids as are generally used in combination with it.

Premising that it has not been forgotten by the student that any plane section of the sphere when viewed at right angles to the plane of section is a *circle*, and when seen at an angle to that plane, an *ellipse*, it is assumed that the intersections of the sphere by any of the plane-surfaced solids will present little difficulty, and therefore will not necessitate more than two or three such problems being given for solution. The intersections of the sphere by solids having curved surfaces, being much more difficult, will require a close attention to the procedure followed in determining them. The first problem, therefore, in this connection is—

Problem 81 (Fig. 185).—*A sphere is penetrated vertically and centrally by a square prism whose sides make equal angles with the VP; it is required to find the lines of penetration of the solids, in plan and elevation.*

Here, it is evident from the way in which the sphere is penetrated, that its axis, and that of the prism, coincide. Therefore, to find the

MECHANICAL AND ENGINEERING DRAWING 179

lines required, first draw in the plan of the solids, with the sides of
the prism at an angle of 45° with the IL or VP, as in No. 1. Then
find by projection the elevation of them as if entire. As the axes
of the sphere and prism are coincident, their lines of intersection in
plan are in the planes of the sides of the prism, and therefore coincide
with the four lines in No. 1 representing its plan. Then, in the
elevation No. 2, as the sides of the prism are *equally* inclined to the
vertical plane of projection—or the VP—the lines of penetration of
the sphere by the prism will all be portions of ellipses, or parts of
vertical plane sections of the sphere—taken through it at the sides
or faces of the prism—seen at an angle of 45°. To draw in these
lines, find from the plan No. 1 the major and minor axes of the
ellipses into which the circular sections of the sphere are projected
in elevation, and by means of the paper trammel—previously explained
—find the points through which the lines of intersection are drawn, as
shown in No. 2 in the figure.

As the lines of intersection of a prism—having *any number* of
sides—with a sphere, are but a series of circular or elliptic arcs, or
both, obtained in the same way as those in the case of the square
prism, further examples of their penetrations are unnecessary; and as
the bounding edges of the faces of a prism are parallel, the next
problem is a variant from this, or one in which the sides and edges
of the penetrating solid *incline* equally to its axis. The problem is—

Problem 82 (Fig. 186).—*A sphere is penetrated by the frustum
of a square pyramid, having its base edges parallel and perpendicular to the VP; required the lines of penetration of the solids, in
plan and elevation, when their axes coincide and are in a vertical
position.*

First draw in an elevation of the frustum of a square pyramid, as
if entire, in the position given in the problem. At about the middle
of its height, and on its axial line, describe a circle to represent the
sphere to be penetrated; then find by projection a plan of the solids,
as in No. 2 in the figure. Now, as only part of the front face of the
frustum is visible in the position given, the lines of penetration on
that face only will be seen in elevation. This face being inclined to
the vertical, it is evident that a section of the sphere taken at it,
although actually *circular*, will become *elliptic* in projection; but in
this case very slightly so, due to the small inclination of the sides of
the frustum.

To draw in the portions of the ellipse, which are in fact the *lines
of penetration* required in elevation, its two axes must first be found.
One of them, so far as its position is concerned, will coincide with the
axis of the frustum; and the other, which will be at right angles to it,
must be drawn through the point which is the actual *centre* of the
circular section of the sphere, made by a plane coinciding with any
one of the sides of the frustum. For convenience, take that side
which is on the right in No. 1, and bisect that part of it which is
contained between the points *a, b*. Through the point of bisection

180   FIRST PRINCIPLES OF

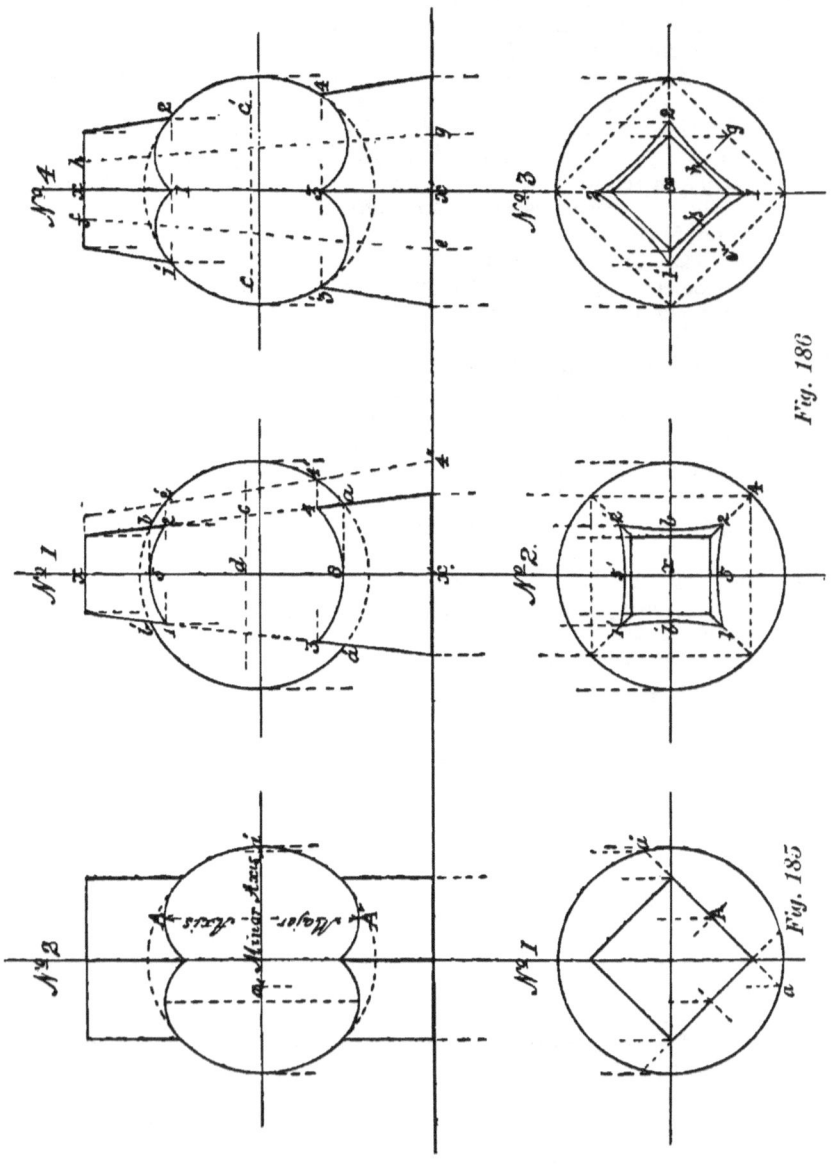

Fig. 185

Fig. 186

MECHANICAL AND ENGINEERING DRAWING   181

c draw a line perpendicular to the axis of the frustum, and on it set off
from d—where the axes intersect—on either side, a distance equal
to c, a or c, b, which is half the major axis of the elliptic section.
For the minor axis, project over to that of the frustum the points a, b
in its right side or edge; then with these lengths, and the paper
trammel, mark off a few points at either end, and through them draw
the elliptic arcs to the right and left edges of the front face of the
frustum, and they will be the lines of penetration of the solids seen
in elevation.

For the plans of these lines on the upper surface of the sphere, let
fall projectors from points 1, 2, in No. 1, to cut the corresponding slant
edges of the frustum (in No. 2) in points 1 1', 2 2'; and for points
$b b'$, 5 5', in the same diagram, set off from x, on the two lines drawn
through it at right angles to each other, the distance that b or b', in
No. 1, is from the axial line—or point 5; then curved lines drawn
through the points thus found, as shown in No. 2, will be the lines
of penetration on the *upper* surface of the sphere by the frustum.
The corresponding lines on its *under* surface, if required, are found in
the same way.   If the sides of the frustum are equally inclined to the
VP, as in the case of the square prism in Fig. 185, and an elevation
of the solids in this position be required, then, as in previous problems,
transfer the plan No. 2 to the required position shown in No. 3, and
from it and the elevation No. 1 find by direct projection the view
given in No. 4.   The dotted lines and projectors show how the major
axes of the elliptical portions of the intersections are obtained, the
line $c\,c'$, in which the minor ones lie, being a projector drawn through
d in No. 1.

78. As a sphere penetrated *centrally* by a cylinder or cone would
in each case give circles as the lines of intersection, becoming in
projection either straight lines, circles, or ellipses, according to their
positions with respect to the planes of projection, it is considered
unnecessary to give any problems for solution with the solids so
combined, as the truth of the statement is self-evident without
illustration.   It must not, however, be forgotten by the student that
such combinations of the sphere with the cylinder or cone, or both,
more frequently occur in practice than any other; as, for instance,
in many kinds of cocks and valves, tee-pieces, handrails, stanchions,
boiler mountings, etc.   As cases, however, arise where these three solids
are in such combination that their axes are *not* coincident, or even in
one plane, the remaining problems in this part of the subject will have
reference to instances in which these infrequent junctions occur.

Problem 83 (Fig. 187).—*A sphere is penetrated by a cylinder whose
axis is vertical and parallel to that of the sphere; required the
lines of intersection of the solids, when their axes are a given
distance apart, in a vertical plane inclined at an angle to the VP.*

Let No. 1 (Fig. 187) be a plan of the solids, the larger circle being
that of the sphere, and the line $v\,p$, drawn through its axis a, and that
of the cylinder c, a plan of a vertical plane at an assumed angle of

182   FIRST PRINCIPLES OF

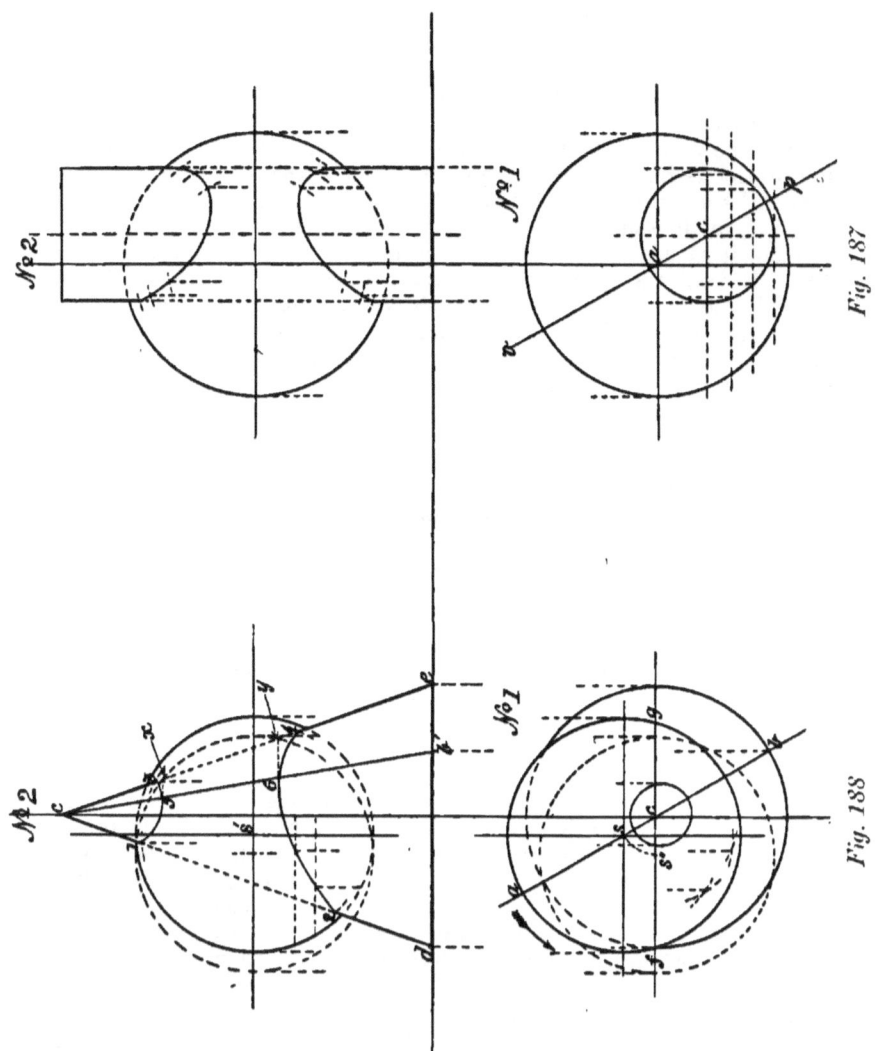

Fig. 187

Fig. 188

MECHANICAL AND ENGINEERING DRAWING   183

60° with the VP. Then as vertical sections of the sphere parallel to the VP will be simply circles, and similar sections of the cylinder, rectangles, it is evident that the visible lines of penetration of the two solids may at once be found by assuming vertical planes to pass simultaneously through them both, and drawing lines through the points of intersection of the sections thus produced, giving the lines required. Therefore, having the plan of the solids given in position, as in No. 1 in the figure, from this obtain by projection an elevation of them as if entire. Then draw in, parallel to the IL and through the front halves of the sphere and cylinder, as many lines as it is intended to use section planes; find by projection elevations of the circular and rectangular sections produced by them, and through the points of intersection of these, draw in the lines of penetration as shown in No. 2, Fig. 187.

The next problem is a combination of a sphere with a cone. It is—

Problem 84 (Fig. 188).—*A sphere is penetrated by a cone having its axis vertical; required the lines of intersection of the solids when their axes are not coincident, but parallel, and lie in a plane which is inclined to the VP.*

It may not have occurred to the student, in dealing with the previous problems in which the sphere is involved, that it is quite possible for its axis to lie in a *vertical* plane and yet be inclined at an angle to either or both of the planes of projection. In the problem for solution the axis of the cone being vertical, and that of the sphere parallel to it, both will be vertical, and be represented by points in plan. Therefore in No. 1 (Fig. 188) let the points c and s be the plans of the axes of the two solids, c being that of the cone and s that of the sphere, both being in a plane—of which the line $a\,b$ is the plan —at an angle of 60° with the VP. Then as the sphere is the solid penetrated, with s as centre, and half the sphere's diameter as radius, describe the circle to represent its plan, and with c as a centre and half the diameter of the base of the intended cone as radius, describe a second circle, dotting in that part of it covered by the sphere, as shown · in No. 1 in the figure. From this plan find by projection the elevation of the two solids as if entire, making the height of the centre of the sphere—from the IL—about half the height of the cone. Then to find the curved lines of penetration the *principal* points in them should first be ascertained.

For points 1, 2, 3, 4, in No. 2, get a sectional elevation of the sphere on the line $f\,g$ in No. 1. The section obtained, which is a circle, having a diameter equal to $f\,g$, will be found to cut the sides of the cone $c\,d$ and $c\,e$, in 1, 2, and 3, 4. For points 5 and 6—the lowest and highest in the curves of intersection in the two hemispheres —conceive the *cone* to be turned *on its axis* in the direction of the arrow, carrying the sphere with it, until the line $a\,b$ in No. 1 becomes parallel to the IL. In so doing, the axis s will have moved to s'', and the sphere itself to the dotted positions shown in No. 1 and

184   FIRST PRINCIPLES OF

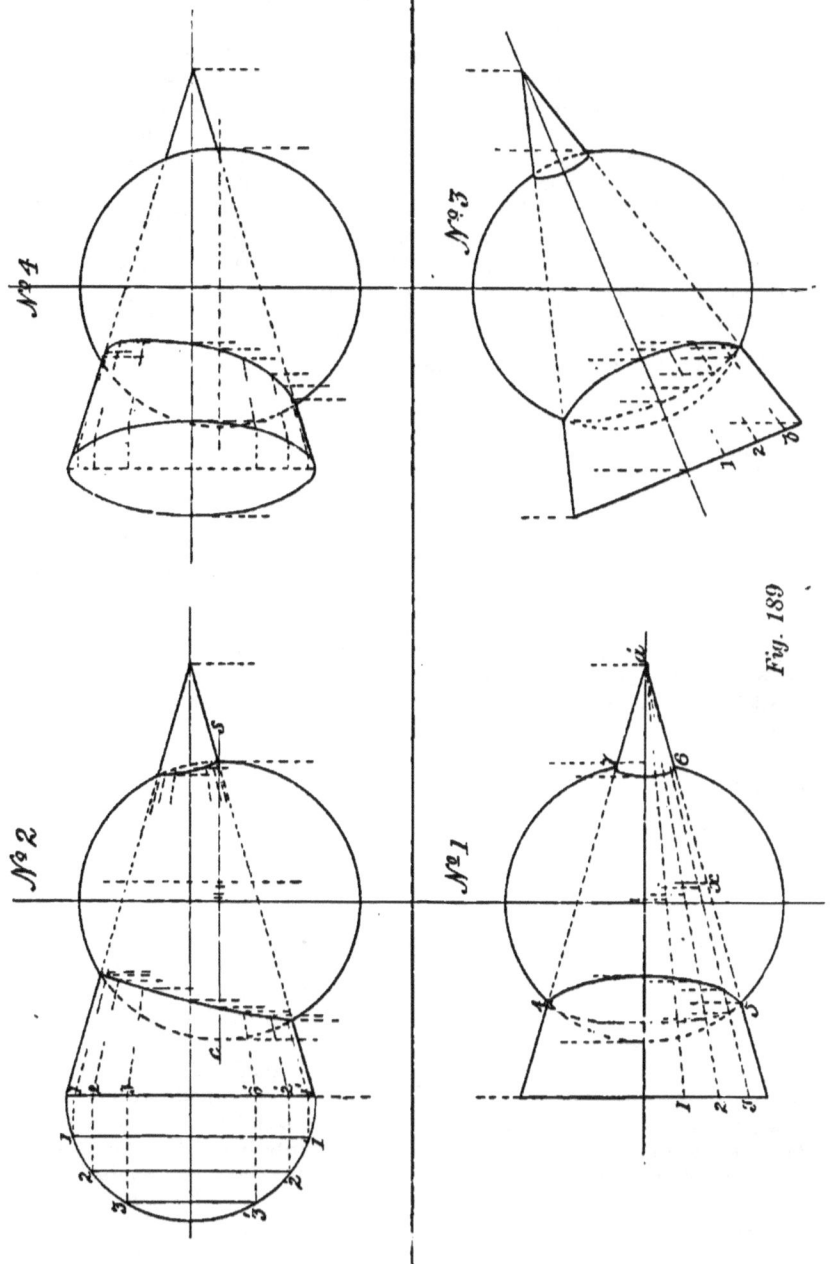

Fig. 189

MECHANICAL AND ENGINEERING DRAWING   185

No. 2, the latter giving points $x\,y$, which will determine the actual positions of the points sought on the conical surface, with the solids, as originally placed. To fix these, find in No. 2 the elevation cb' of the line from c to b in No. 1, and to it project over—parallel to the IL— the points $(x\,y)$ last found, cutting it in 5 and 6, which are those required, the projecting over of x and y being tantamount to turning the cone back on its axis to its original position. For other points in the lines of penetration, take a few horizontal sections—which will all be circles—of both solids, between those already found, and through their points of intersection, in plan and elevation, draw lines, and they will be the lines of intersection of the surfaces of the cone and sphere, as required in the problem, and shown in No. 1 and No. 2 (Fig. 188).

79. The next problem will show that although sections of the two solids might be taken in such a way that they would produce circles for both cone and sphere, and give apparently an easy solution of it, it is necessary for the draughtsman to exercise judgment in deciding upon his method of procedure. The problem is—

Problem 85 (Fig. 189).—*A sphere is penetrated by a cone horizontally; required the lines of penetration in plan and elevation, when the axes of both solids are in a plane parallel to the VP, and at right angles to each other, but that of the cone not passing through the centre of the sphere.*

Draw in first, as if entire, the plan of a sphere penetrated by a cone, having its axis parallel to the IL, and passing through that of the sphere, as shown in No. 1 (Fig. 189). From this plan get an elevation of the solids, showing the axis of the cone some distance *above* the centre of the sphere. Now, if sections of the solids be taken by planes *parallel* to the axis of the *sphere*—which is assumed to be vertical—and at right angles to the VP, the sections would in reality be circles for both solids, but only straight lines in projection in both plan and elevation, which would be useless in determining their intersections.

Again, sections taken through both solids parallel to the axis of the *cone*, although giving circles for the sphere, would still give hyperbolas for the cone, which are not desirable. The simplest method in solving the problem will therefore be to assume vertical section planes to be taken through the apex of the cone in varying directions and passing through both solids simultaneously. These would give triangles for the cone and circles for the sphere; the latter, however, being at an angle to the VP, would in elevation be projected into ellipses. Therefore in No. 1 (Fig. 189) draw in, through a' the vertex of the cone, to its base, as many lines—say three—$a'1\ a'2\ a'3$, as it is intended to use section planes. Next find in No. 2 the elevation of the triangular sections produced by these planes, as in previous problems. Then, so far as the sections of the sphere are concerned, all that is required is to find the *points* in the ellipses—into which the circular sections of the sphere are projected—which intersect the

MECHANICAL AND ENGINEERING DRAWING

triangular sections of the cone, for it is through them that the lines of penetration of the sphere by the cone will pass.

To determine these points, there is no necessity to draw in the elliptic projections, as they may be found very readily by the use of the paper trammel before mentioned. Having numbered the section planes, in plan and elevation, as in No. 1 and No. 2 in the figure, and noted the points 4, 5, 6, 7, where they cut the great circle of the sphere in No. 1, draw in, in No. 2 through line $c s$, which passes through the centre of the sphere, the lines in which the *major* axes of the ellipses will lie. One of these only—viz., that from x in No. 1—is drawn in, in the diagram, so as to prevent confusion. Then with half the major and projected half of the minor axes of each ellipse—there being one for each section taken—on the trammel, manipulated as explained, *all* the points in the lines of intersection, except the extreme ones, are found as shown in elevation No. 2. For the plans of these lines that will be seen from above—shown in No. 1—let fall projectors from the points in the lines 1, 2, 3, 4, in No. 2, to cut the corresponding ones in No. 1, and through the intersections draw the curved lines as shown, which will be the ones required.

Should the axis of the cone be inclined to the VP, still remaining parallel to the HP, and the lines of intersection of the two solids be required, when in this position, proceed as in previous cases to transfer the plan No. 1 to such a position in No. 3 as will bring the axis of the cone to make the required angle with the IL, and then from it and the elevation No. 2 obtain by direct projection the view shown in No. 4.

From the problems which have been given in this part of the subject, the student will be able to appreciate the endless variety of positions the solids in combination may be made to assume, but as it is not our object to multiply examples, but to give only such combinations as are likely to occur in practice, we pass on now to the last section in the application of the principles of projection—viz., the "Development of the Surfaces of Solids," a subject of the highest importance, not only to the draughtsman, but to all who have to shape or fashion anything constructed of sheet metal.

CHAPTER XX

THE DEVELOPMENT OF THE SURFACES OF SOLIDS

80. To determine the exact form of the surface of any solid, whether it be plane or curved, it is necessary to obtain what is known as its "development"; or, in other words, the particular shape its surface will assume when laid out *flat*, supposing it possible that it can be so treated.

For such a surface to be "developable" it must be one, on every part of which a sheet of any flexible—but non-elastic—material can be made to lie when bent, without leaving any hollow spaces. Should this not be possible, then the surface is non-developable.

From this definition, it will at once be seen, that all *plane*-surfaced solids are developable, while of those having curved surfaces, only the cylinder, and the cone, with their frustums, fall within the same category; the sphere, with the spheroids, ellipsoids, and many other solids of revolution, having surfaces which will not coincide with a plane when laid out flat, but would tear or crease, being non-developable.

The figure of the developed *surface* of every solid, which when bent will cover it at any and every point, is called the "envelope" of that surface.

In the cylinder and cone, as well as in every other solid of revolution, any line drawn on their surfaces, in the same plane as their axes, is called a "meridian." If such a line is straight, the surface is developable, but if curved it is non-developable.

A surface which is generated by the motion of a straight line is called a "ruled" surface, and may be either developable or non-developable; if the latter, it is a "twisted" surface, in that it cannot be laid out on a plane without being torn. A ruled surface may, however, be curved, and developable, and yet form no part of a cylinder or a cone, as will be shown later on.

The finding of the developments of plane-surfaced solids involving no difficulty, few problems are necessary in connection with them, as it will be seldom that any will occur in the practice of the student, with which he will not be able successfully to grapple. The first is the simplest possible, and hardly requires demonstration, but as it shows

187

188　　　　　　　FIRST PRINCIPLES OF

the method of procedure in finding the development of the surface of any plane solid, it is here given.

Problem 86. (Fig. 190).—*Given the plan of a cube; to find the development of its surface.*

Let *a b c d* in the diagram be the plan of the cube. Then as its six sides are all of them equal squares, with every two adjacent ones at right angles, all that is required in finding its development, is to conceive each of its vertical faces turned down on its lower edge as a hinge, until it lays flat on the HP, one of the faces in its turning carrying the top face *a b c d* with it.

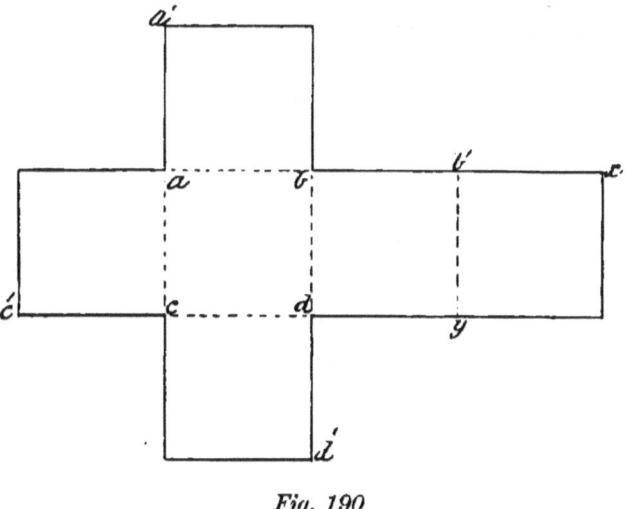

Fig. 190

To show this graphically, in Fig. 190, produce the four sides of the square *a b c d* indefinitely; then from its four corners, set off on the produced lines at *a'*, *b'*, *c'*, *d'*, a length equal to any one side of the cube, and through the points thus found, draw lines parallel to its sides, as shown. From *b'* in *a b* produced, set off a length *b'x*, equal to that of the edge *a b*, and at *x* draw a line parallel to *b'y*, and the development of the cube will be complete. The surface enclosed within the bounding lines of the diagram will then be the "envelope" of the given solid, for it is its whole surface laid out *flat*, and will, if cut out of a sheet of paper, and folded over on the lines *a c*, *c d*, *d b*, *b a*, and *b y*, until each adjoining surface is at right angles, exactly cover all the cube, without leaving any vacant space between it and them.

81. As the development of the *whole* of the surface of a square prism would merely be a repetition of that given of the cube, in the

MECHANICAL AND ENGINEERING DRAWING

next problem only a part of a prism is taken, as the solid whose development is required.

Problem 87 (Fig. 191).—*Given the elevation of a truncated square prism, to find the development of its surface.*

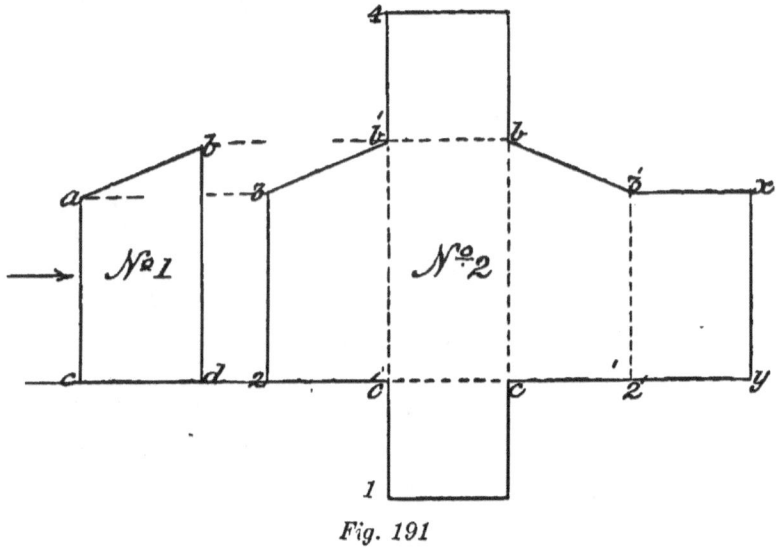

Fig. 191

Let No. 1 (Fig. 191) be the side elevation of the prism, standing on its base in a vertical position. As its base is square, its vertical sides will all be of the same width; therefore on the line cd produced indefinitely, draw in first an elevation of the prism, looking at it in the direction of the arrow. From c and c' in it, set off in points 2, 2' a distance equal to cd in No. 1, and through them draw vertical lines, and cut them in points 3 and 3', by a projector from a. Join b' 3 and b 3'. For the base, and top surface of the prism, produce the lines $c'b'$, cb, in both directions indefinitely. Then with c' as centre, and $c'2$ as radius, cut $b'c'$ produced in point 1; and from b', with b' 3 as radius, cut the same line in point 4, and complete the rectangle 4b (above) and the square $c'1$ (below) as shown. From point 2 in cd produced set off a length 2'y, equal to $c2'$, and through point y draw yx, parallel to 2' 3'; join 3' and x, and the development is completed. If the resultant figure (the envelope of the prism) be cut out in paper, and treated in the same way as that of the cube, it will be found to fit the solid exactly.

82. As the surface of an oblique prism is often found to be more difficult of development by the student draughtsman than that of a right one, the next problem will show how it may be correctly found.

190   FIRST PRINCIPLES OF

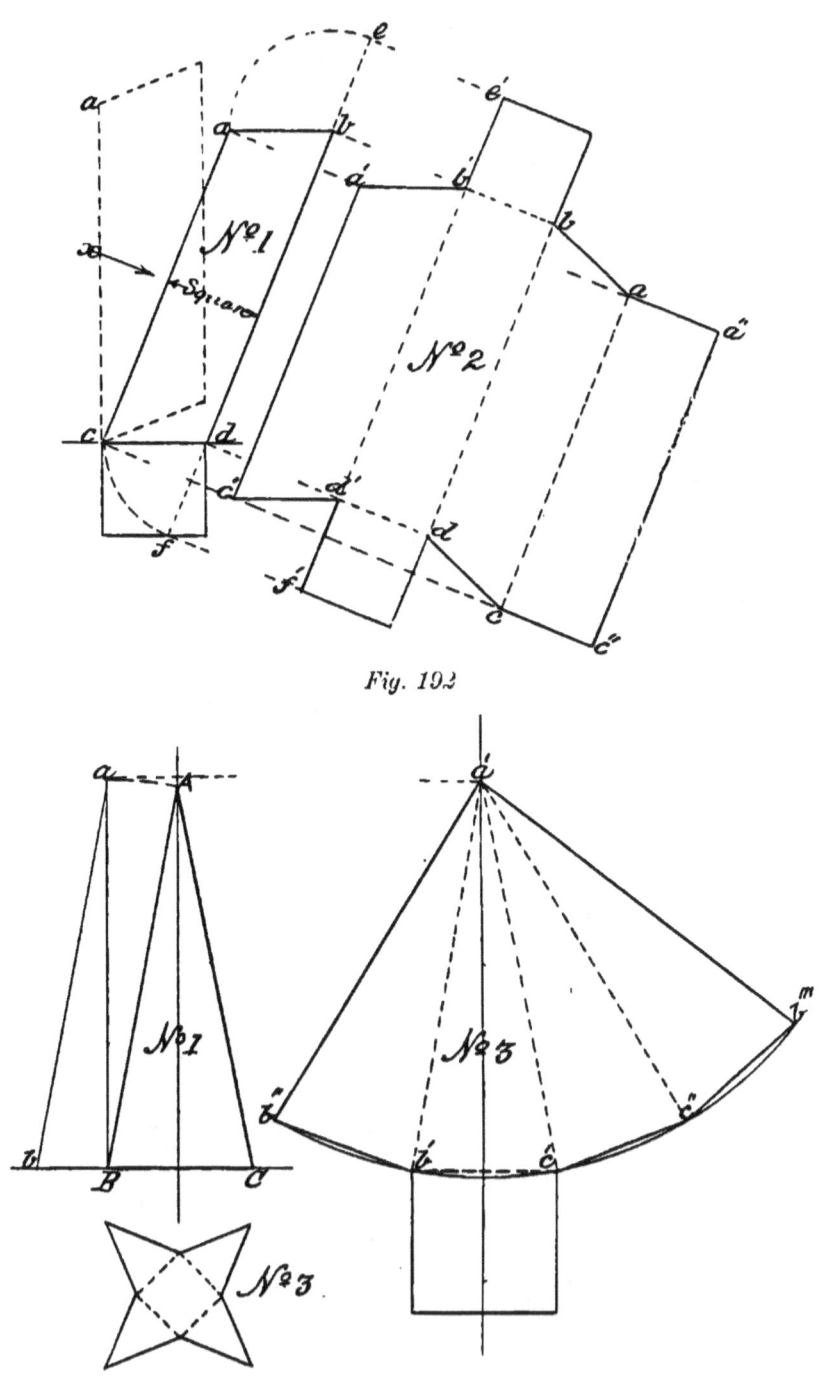

Fig. 192

MECHANICAL AND ENGINEERING DRAWING 191

Problem 88 (Fig. 192).—*Given the side elevation of an oblique prism; to find the development of its surface.*

Let *a b c d*, No. 1, Fig. 192, be the given elevation of the prism, with its end *c d* resting on a horizontal plane. In this position, it is evident that an ordinary front elevation of it, as of that of the frustum in No. 2, Fig. 191, would be of no service in this case, as the bounding edges of its sides, being *inclined* to the horizontal, would not give *actual* but only apparent lengths in projection. To find the actual sizes of all the sides and ends of the solid, and their relative position to each other, on a flat surface, a view directly at right angles to one of the sides of the prism is necessary.

Now this view may be found in two ways. The prism may be turned on its horizontal edge at *c*, as on a hinge, until its inclined edge *a c* becomes vertical—as shown in dotted lines—and its development found when so posed. Or, it may be found at once from the prism in its inclined position, with less chances of error, by a projection of its bounding surfaces, taken when looked at in the direction of the arrow *x ;* such a view being tantamount to assuming the IL of the plane of projection to be drawn through *c*, at right angles to the edge *c a* of the prism, and its surface laid out flat on the VP. To find the development as shown in No. 2, Fig. 192, proceed as follows :—At right angles to *a c*—in No. 1 in the figure—and through *c*, draw a line indefinitely. At any convenient point, as *c'* in that line, draw through *c'* a line parallel to *c a* in No. 1. Then as all the side edges of the prism are parallel to each other, set off on the line drawn through *c* in No. 1 from *c'*, the distances—measured at right angles—that those edges are apart ; and through the points in *c c'* produced thus found, draw lines indefinitely, parallel to that first drawn from *c'*, or to *c a* No. 1.

Now it is evident that the four side (or inclined) edges of the prism will, when the surface is unfolded, lie in the lines last drawn, for if the prism No. 1 as a solid, is laid with its edge *ca* coinciding with the line *c'a'* in No. 2, and rolled over—to the right—its edges would fall upon the lines drawn parallel to *c'a'*.

To complete the development, project over from No. 1—at right angles to *a c*—the points *a c ; b d ;* to cut the corresponding edges in No. 2 ; in *a'c', a c, a"c" ;* and *b'd', bd*. Join *a'b', b a, a a" ;* and *c'd', d c, c c"*, by straight lines as shown. For the two *ends* of the prism—which is square in cross-section—produce the line *b d* in No. 1 in both directions, and from *b* and *d*, set off in *e* and *f*, a length equal to *a b* or *c d*. Project over *e, f* to *e'f'* in No. 2, and through them draw lines at right angles to *b'd"*, to cut *b d*, which completes the development required. The envelope of the prism thus produced will, when folded over on the edges represented in dotted lines, be found to cover without vacuities all the surfaces of the given prism.

83. As the development of the surface of any prism—whether right or oblique—having any number of sides, may be found as above shown, a pyramid is taken as the next object for its surface development. The problem is—

192   FIRST PRINCIPLES OF

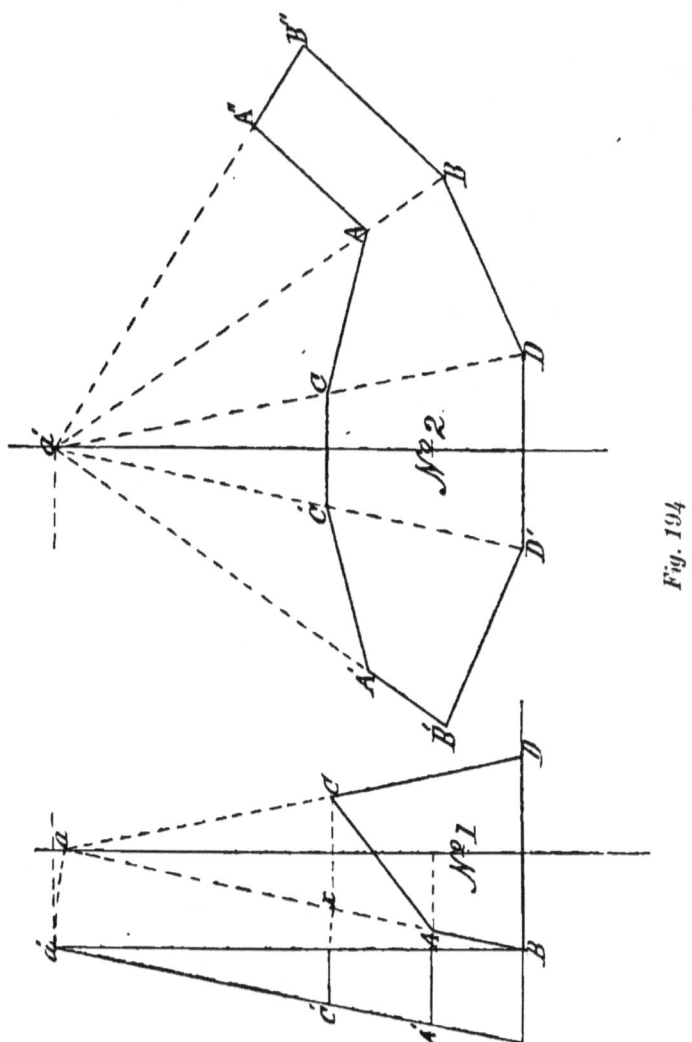

Fig. 194

MECHANICAL AND ENGINEERING DRAWING   193

Problem 89 (Fig. 193).—*Given the elevation of a square pyramid; to find the development of its surface.*

Let ABC, No. 1, in Fig. 193, be the elevation of the pyramid. As all its sides are alike, and incline equally to its apex, it is first necessary to know the *actual* length of one of its side edges. To find this, at B in No. 1, draw a line perpendicular to the IL. With B as centre, and BA—the length of a side of the pyramid—as radius, describe an arc cutting the perpendicular line in a. From B, set off in b, a length equal to half BC, and join a, b; then aB will be the length of a *side* of the pyramid, and $a b$ that of one of its edges.

For the development, at a convenient distance from No. 1, draw a line perpendicular to the IL, and project over to it in a' the point a in No. 1. Then with a' as centre, and $a b$ in No. 1 as radius, describe an arc indefinitely, which will cut the IL in points $b'c'$. From b', set off— on the arc—in b'', a length equal to $b'c'$, and from c' the same length twice in c'' and b''; join $b'b'''$; $c'c''$; $c''b'''$; and draw lines from a' to these points. For the base, construct a square on $b'c'$ as shown, and the development is complete; the figure enclosed within the boundary lines of No. 2 being the envelope of the given pyramid.

A development of the same surface of the form shown in the small diagram No. 3, may also be found by turning down the four sides of the pyramid on their respective base edges. If, however, such an object—if large—were made in metal plate, this development would involve a great waste of material, and necessitate more seams than are required.

84. As the frustum of a hollow pyramid is often combined with parts of other solids in metal plate constructions, a problem in finding the development of such a surface is next given.

Problem 90 (Fig. 194).—*Given the elevation of the frustum of a hollow square pyramid; to find the development of its surface.*

Let ABCD, No. 1, Fig. 194, be the elevation of the frustum. Produce its edges AB and CD, till they meet in a, and find the actual length of one side, and an edge of the pyramid of which it is a part, as in the last problem. Then draw in, as in No. 2, the development of the base edges of the frustum. For its top edges, through A and C in No. 1, draw faint lines across the front face of the pyramid, parallel to its base BD, and with arcs struck from B as centre, transfer the points A and x in Ba to the vertical line Ba', and project them over to C'A' in the line $a'b$. Then from a' in No. 2, set off on the corresponding lines of the pyramid, the distances that C', A', are from a' in the line $a'b$ in No. 1; join the points thus found by straight lines as shown in No. 2, and they will be the development of the edge required. By connecting A'B'; A''B''—at the ends of the diagram—by lines as shown, the development of the whole surface of the hollow frustum will be completed.

85. As solids of a pyramidal form, with *curved*, in place of plane

o

194   FIRST PRINCIPLES OF

sides, are often used in giving shape to mechanical details, the solution
of the following problem will show how the development of the surfaces
of such solids is found.

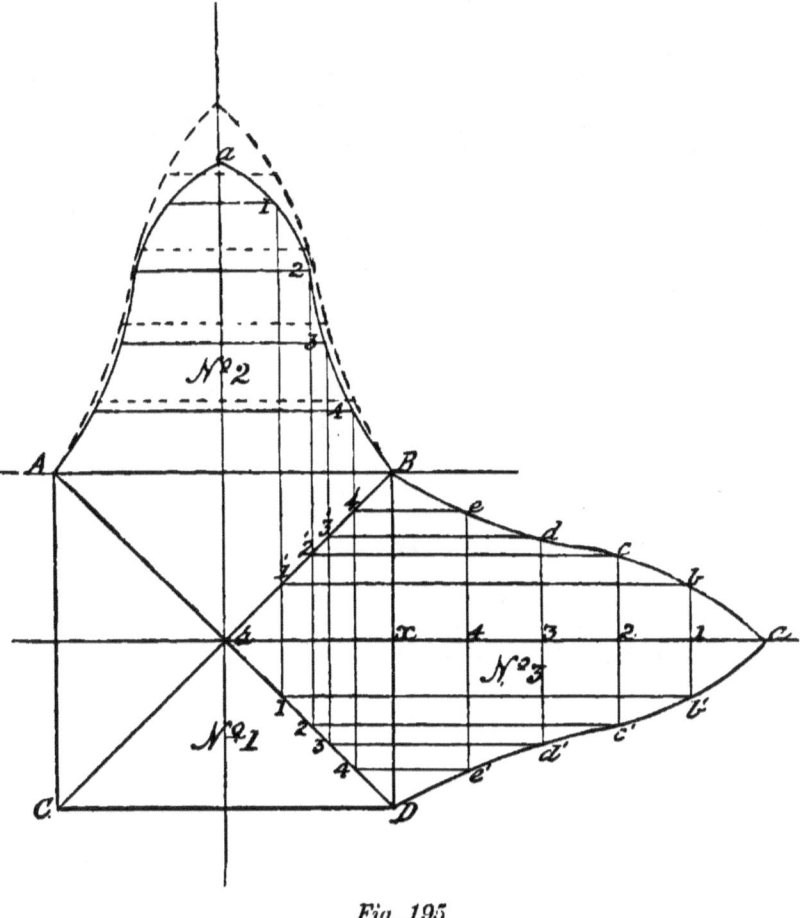

Fig. 195

Problem 91 (Fig. 195).—*Given the plan and elevation of a square
pyramidal-shaped solid ; to find the development of its surface.*

Let ABCD, No. 1, Fig. 195, be the plan, and A*a*B, No. 2, the
elevation of the solid, its axis *aa'* being perpendicular to its base.
Divide the side *a*B, No. 2, into any number—say five—of equal parts,
and through 1, 2, 3, 4, the points of division, draw lines parallel to
AB and BD. Through *a'*—the axis—in No. 1, draw a line indefinitely,
cutting BD in *x ;* from *x* set off in this line the equal parts that *a*B,

MECHANICAL AND ENGINEERING DRAWING   195

No. 2, is divided into, and through the points of division draw faint
lines parallel to BD.   Then the intersection of these, by lines drawn—
parallel to AB—through the points in the diagonals BC and AD, cut
by those let fall from 1, 2, 3, 4, in αB, No. 2, will give *b, c, d, e ;* and
b′, c′, d′, e′; through which, from B and D to *a*, the curved lines shown
are drawn.   The surface enclosed by them and the line BD will be the
development of one side of the given solid.   As its axis is perpendicular
to its base, and all its faces are alike, the development found of one
face, if repeated on the other base edges (AB, AC, CD) of the solid,
will, with the base itself, give the complete development required.   The
difference between the apparent and real surface of a side of the solid
is shown in No. 2 in the diagram, where the full lines give the apparent
surface when bent to its shape, and the dotted ones the same surface
laid out flat.

The solid here dealt with, has been chosen to show that although it
has compound curved surfaces—convex and concave—combined with a
flat one—its base—yet it can all be made, if necessary, though at a
great waste of material, out of a single flat plate, its sides being afterwards bent to the shape required.

86. As the whole, or a part of the surface of another plane solid,
the "oblique pyramid" (which often contributes in giving form to plate
and other metal structures), is rather more difficult of development than
that of a "right pyramid"—on account of its frequent great inclination
from the vertical—the solution of the next problem will show how its
correct development may be found with the least number of construction lines.

Problem 92 (Fig. 196).—*Given the elevation of an oblique square
pyramid, to find the development of its side surfaces, when its axis
is inclined 45° from the vertical; also the development of the surface
of a frustum of the same pyramid, of a given vertical height.*

Let ABC, No. 1 (Fig. 196), be the elevation of the pyramid.   As all
its sides with their edges are inclined to both the VP and HP, it is
evident that the actual shape of the whole or any part of its surface,
cannot be found by direct projection, or from the elevation alone,
without a plan of the solid.   Now although an oblique pyramid differs
from a right one, in having its axis inclined to its base it must be
remembered that any section of it parallel to its base is still—as in the
case of a right pyramid—of the same *form* as its base.   Bearing this in
mind, there will be no difficulty in solving the given problem, more
particularly the latter part of it.

The elevation of the pyramid being given, find by projection a plan
of it, as shown in full lines in No. 2, Fig. 196, keeping the axial line
A*a* a convenient distance from its base BC, in No. 1.   Then to find the
development of the surface required: in No. 3, Fig. 196, draw a line
(indefinitely) at right angles to BC the base, and a projector from its
apex parallel to that base, to cut it in A′.   On the line through A′ set
off from A′, the length of the longest side AB of the pyramid in *a′*,
and through *a′* draw a line at right angles to A′*a′*.   From *a′*, set off to

196   FIRST PRINCIPLES OF

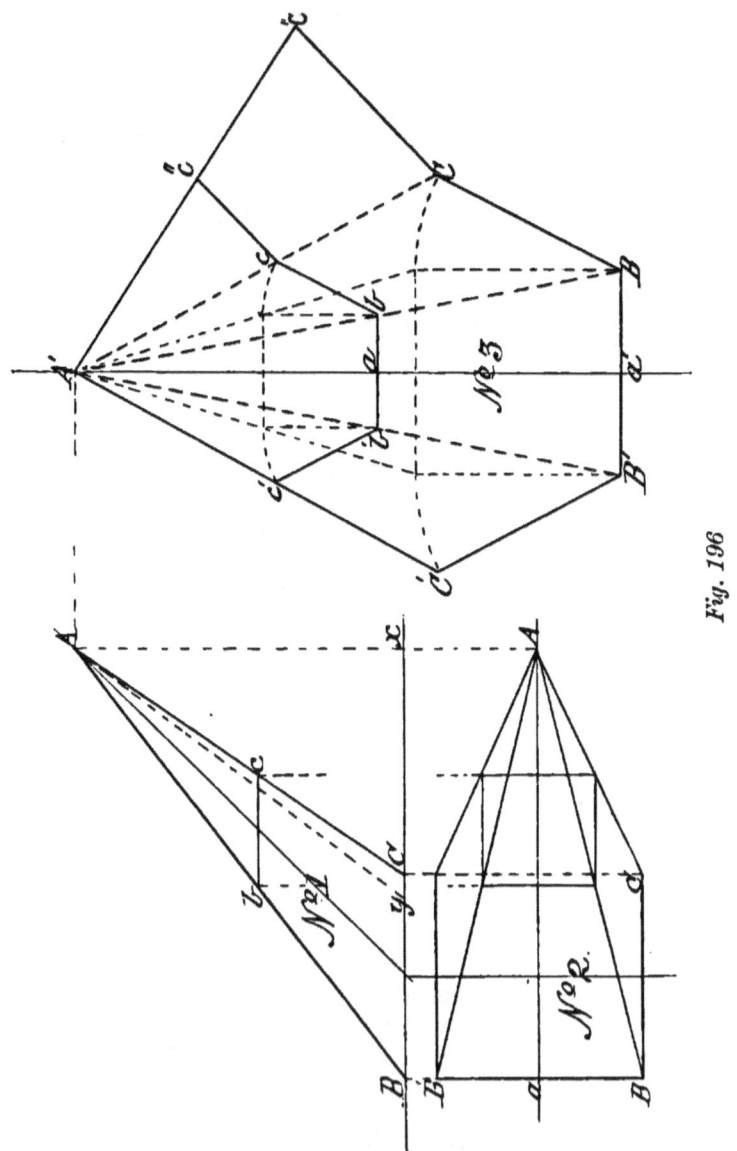

Fig. 196

MECHANICAL AND ENGINEERING DRAWING   197

right and left in B and B′, half the length of one edge of the base of the pyramid, then the line B′B will be the development of the edge B′B in No. 2. Join B and B′ with A′ by straight lines, then will the triangle A′B′B be the development of the longest face of the pyramid.

For the adjacent faces BA′C, and B′A′C′ : with BB′ as radius and B,B′ as centres, describe arcs to right and left, and from A′ as centre with the actual length of the edge AC in No. 1 as radius, cut the arcs last drawn in C and C′. Join A′C′ and A′C by right lines, and the triangles A′C′B′ and A′BC will be the development of the front and back faces of the pyramid. For the shortest face—or that to the right in No. 1— of the pyramid in No. 3, with C as centre, and CB as radius, describe an arc, and from A′ as centre, with AC as radius, cut that arc in C′ ; join A′C″ with a right line, and the development of the four faces of the pyramid when laid out flat is complete.

The actual length of A′C in No. 3 is the *hypothenuse* of a right-angled triangle, of which the line AC in No. 2 is the *base*, and the vertical height of the pyramid—or Ax in No. 1—the *perpendicular*. This hypothenuse is shown by a dotted line drawn from A to y in No. 1, the length xy being equal to AC in No. 2.

For the development of the frustum BbcC of the pyramid, as its top edges are parallel to the base BC, it is only necessary to set off the length Ab in No. 1 from A′ in No. 3, in the point a, and through it draw a line parallel to B′B, cutting A′B′, and A′B in $b′$ and b ; then parallel to BC; B′C′; and CC′; draw b c, $b′c′$, and $c′c″$, and the development is complete. The dotted arcs in No. 3 show the direction the developed side and end surfaces would move in to form the covering of the solids.

87. The foregoing examples of the development of the surfaces of plane solids being sufficient to show the principle on which they are obtained, we pass on to the consideration of those which are bounded by developable *curved* surfaces, such as the cylinder and cone, with their frustums.

Problem 93 (Fig. 197).—*Given the plan and elevation of a right cylinder; to find the development of its curved surface, and that of a given point, and line on that surface.*

Let the circle No. 1, Fig. 197, be the plan of the cylinder, the rectangle bcde, No. 2, its elevation, and the line a $a′$ its axis. Also let x and y be two given points in its front surface. Now, a cylinder has been defined to be a solid generated by the revolution of a rectangle about one of its sides as an axis; or its surface may be conceived to be generated by the motion of a straight line around another which is fixed; the former being always parallel to the latter, and at a given distance from it, thereby causing every point in it to lie in the cylindrical surface throughout its motion. This surface being generated in one complete revolution, it is evident that it may be developed or laid out flat, by causing, as it were, the solid to give up its surface while rolling on a plane through one revolution.

198   FIRST PRINCIPLES OF

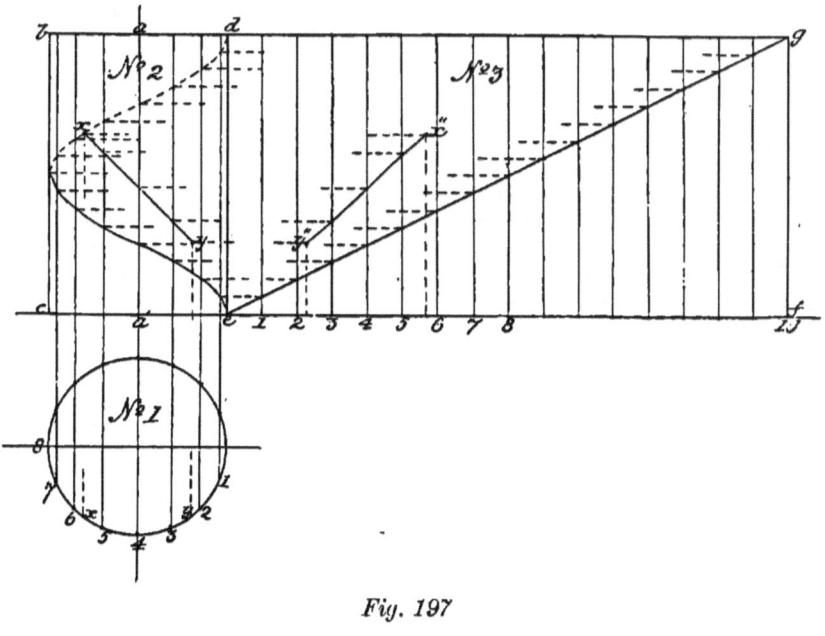

Fig. 197

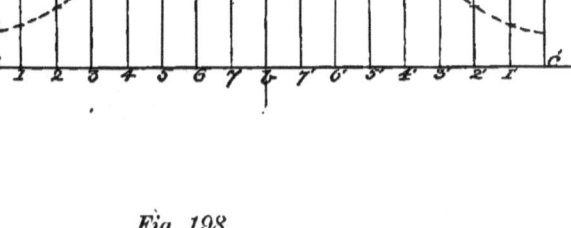

Fig. 198

MECHANICAL AND ENGINEERING DRAWING   199

To show this graphically : in No. 2, Fig. 197, produce indefinitely in one direction—say to the right—parallel to each other, the lines bd and ce, then divide half the circle in No. 1 into any number of equal parts— say eight—and number them as shown. Project over each point thus found to ce No. 2, and through them draw meridians parallel to bc. On ce in No. 2 produced, set off from e the eight equal parts into which the semi-circle 8, 4, 0 is divided, number them as shown, and repeat the divisions to point f. Through points 1 to 8, and f, draw lines parallel to the side de of the cylinder; then will the rectangle $defg$ be the development of the surface of the cylinder $bcde$; for its length ef or dg, equals the circumference, and its width de or gf, the height of the cylinder; and if it be conceived to roll on a plane with that part of its surface represented by the line de in No. 2 touching it when starting, it is evident that it will roll over the surface included in the rectangle $defg$, and at the completion of one revolution, the line de (if drawn on the cylinder's surface) would be found to fall exactly on the line gf, as the meridians 1 to 16—assumed to be drawn on that surface—would each fall on its corresponding one drawn on the development.

By laying out the surface of any solid of revolution (if developable) in a similar way to that of the cylinder, it will be apparent that the development of any point or line on its surface in any position, or direction, may be soon found by the help of a few meridians.

For instance, let xy be two *points* on the front surface of the cylinder, and it is wished to find their position on its envelope. First let fall projectors from x and y in No. 2, on to the plan of the front face of the cylinder in No. 1; they will be found to cut it in $x'y'$; the first between points 5 and 6; and y between 2 and 3 in the semi-circle 8, 4, 0. Then on ef (the developed edge of the base of the cylinder) set off from points 3 and 5 in it, the distances that $x'y'$ are from the corresponding points in the semi-circle 8, 4, 0; and through these draw faint lines parallel to de. Then the points x'',y'' in No. 3, where these lines are cut by projectors drawn through x and y in No. 2 parallel to ef, are the positions of those points on the envelope of the cylinder.

Again, if a *line* be given in position on the surface of a cylinder, drawn, say between the points x and y, and its development is required, then, as a line is the path of a moving point, take any convenient points in the given line xy, and draw meridians through them. Proceed, as in the case of the points x, y, to find the position of these meridians in No. 3— the development of the cylinder—and then through the points chosen in the given line xy in No. 2, draw projectors to cut the new meridians found in No. 3. A line drawn through the points of these intersections will be the development required.

88. Assuming the converse of the last problem, or one where the development of a line is *given*, and its position and delineation on the solid itself is *required*, then to show these, we have in effect to re-cover the solid, with the envelope that contains the given line. Let the problem be—

Problem 94 (Fig. 197).—*Given a straight line e g, drawn on the envelope d e f g of the cylinder ; required the projection of that line on the solid itself.*

To find this, divide the length of the given envelope into any number—say 16—of equal parts, and assume the circle of No. 1, and the rectangle *d e f g* of No. 2, Fig. 197, to be the plan and elevation of the cylinder whose envelope is given. Divide the circumferential line of No. 1 into the same number of equal parts, that the length of the envelope is divided into, and through the points of division draw meridians as in No. 2. On to each of these meridians in consecutive order, project over the points where the given line *e g* cuts its corresponding meridian drawn in No. 3. Eight of these—from 1 to 8—will be on the *front* side of the cylinder in No. 2, the remaining eight being on its rear side.

Now, if through the points in the meridians—1 to 8 in No. 2—projected over from No. 3, a line be drawn, it will be that part of the projected line required, which will be on the front side of the cylinder. For the part on its rear side, project over the remaining eight points— 9 to 16—in the line *e g* in No. 3, on to the corresponding meridians in No. 2, and through them draw a dotted line—it being out of sight—and the spiral line thus produced will be the projection of the straight line *e g* in No. 3, when the surface on which it is drawn is wrapped round the cylindrical solid represented in elevation by No. 2, Fig. 197.

In the same way as above shown, may be found—by projection— the delineation on the solid itself of any line drawn on its envelope, the method of doing it being merely the converse of that adopted in finding the development of a given line drawn on a solid.

89. As frustums of a cylinder constantly occur in pipe connections, when it is desirable to alter the direction, or lead of a pipe, the solution of the next problem will show how a plate has to be shaped to form such a connection.

Problem 95 (Fig. 198).—*Given the elevation of the frustum of a right cylinder, the plane of section being at an angle with the axis of the solid ; to find the development of the surface of the frustum.*

Let *a b c d*, No. 1, Fig. 198, be the elevation of the frustum. Here, as the plane of section *a d* of the cylinder is at right angles to the VP, the height of any point in the edge of the section nearest the VP will be the same as that of a point directly in front of it, in the opposite edge ; it is therefore only necessary to give a plan of the front half of the frustum, to enable the whole of its development to be found. Consequently, on the base *b c* of the frustum as a diameter, describe a semi-circle and divide it into any number of equal parts—say eight— and through the points of division draw in the meridians on the frustum. Next produce its base line *b c* indefinitely, and set off on it from *c*, the eight equal parts into which the semi-circle on *b c* is divided. Through the points of division, draw lines parallel to the sides of the

MECHANICAL AND ENGINEERING DRAWING 201

frustum, and number them as shown. Then from each of the points in the line ad, where the meridians drawn on the front side of the frustum intersect it, draw projectors parallel to bc produced, to cut its corresponding meridian in No. 3. A line drawn through the points of intersection of the projectors and meridians, will then give the development of the front half of the frustum.

For the back half, produce the projectors drawn through the points 1 to 7 in ad, to cut the corresponding meridians on the back face of the frustum; then a line drawn through their intersections as shown in No. 2, will give the development of the top edge of the frustum. For the side and bottom edges, draw in, in full, the lines dc, $d'c'$, cc', and the required development is complete.

90. As parts of the surface of an *oblique* cylinder sometimes enter into the design of boiler flues, uptakes, air shafts, etc., it is necessary that the difference between its development and that of a right cylinder should be understood by the student.

The particular difference between the two solids, the right and the oblique cylinder, is, that in the former its axis is perpendicular to its bases or ends; whereas in the latter it is inclined to them; the ends in both cases being circular and parallel to each other. There is, however, another important difference, which affects their developments; viz., that a section of either, taken parallel to its ends, is a circle; while a section at right angles to the axis of either is a circle for the *right* cylinder, and an ellipse for the *oblique* one. This difference in cross-section, it may interest the student to know, is the reason why all circular vessels or pipes intended to withstand an *internal* pressure when in use, are right cylinders, while those employed as mere conduits for the passage of air, smoke, or light gases, not under pressure, may be made in whole or in part of oblique cylinders. The solution of the following problem will show how the development of the surface of an oblique cylinder is found.

Problem 96 (Fig. 199).—*To find the development of the surface of an oblique cylinder of a given diameter and length, and having its axis inclined at a given angle.*

Let the inclination of the axis of the cylinder be 60°; then to find the development of its surface, we must first draw its elevation. Assuming it to be resting on one of its ends on a horizontal plane, take the IL as that plane, and at any convenient point in it, as a in No. 1, Fig. 199, draw a line making with the IL an angle of 60°. With a as centre, and half of the intended diameter of the base of the cylinder as radius, describe a semi-circle cutting the IL in b and c; and through those points draw lines parallel to aa'. On the axial line, set off from a the intended length of the cylinder, and through the point thus given, draw the line dc parallel to bc; then $dbce$ will be the elevation of the cylinder.

Next, divide the semi-circle into any number—say eight—of equal parts, and through each point of division draw a line—perpendicular to bc—to cut bc in the points 1, 2, 3, etc.; and through these draw

202   FIRST PRINCIPLES OF

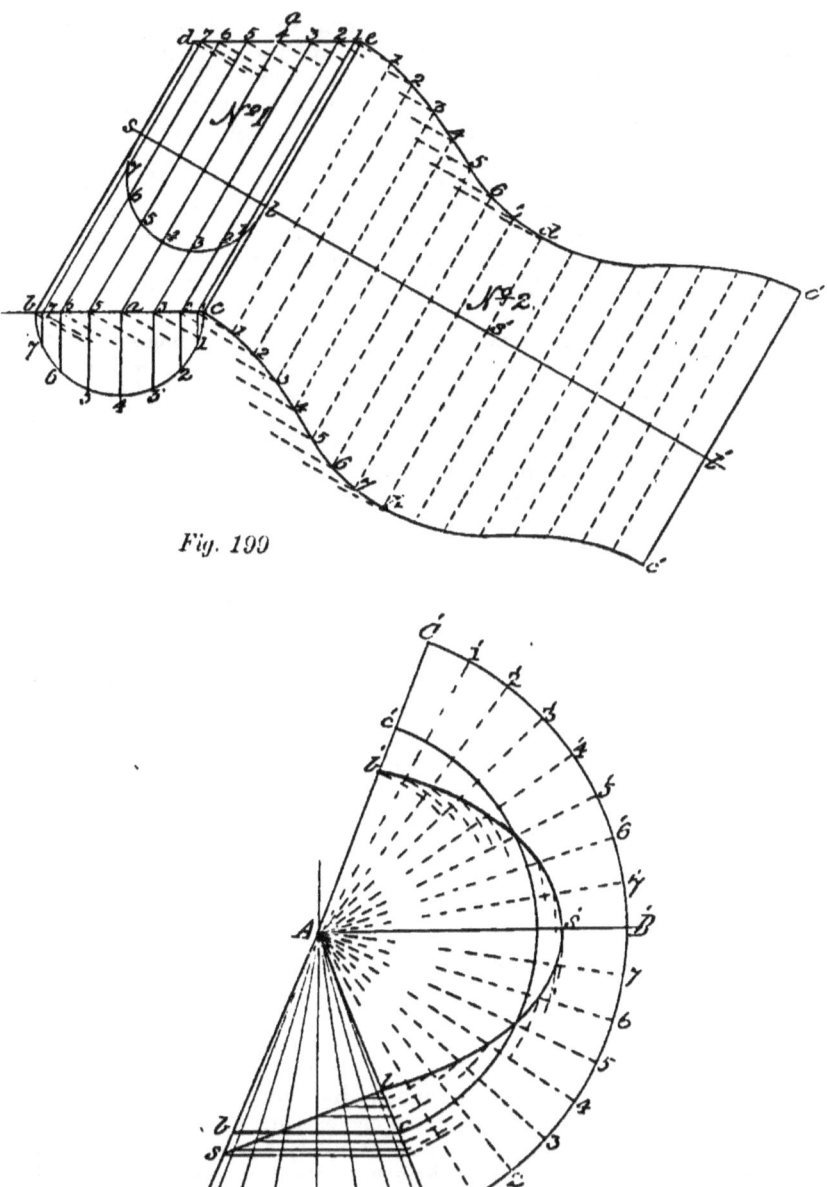

Fig. 199

MECHANICAL AND ENGINEERING DRAWING   203

faint lines parallel to the axis $a\,a'$. At any point, s in $b\,d$, draw the line $s\,l$ at right angles to $a\,a'$, and produce it indefinitely. From the points in $s\,l$ where the meridians 1, 2, 3, etc., cross it, set off on each of these respectively, the lengths of the ordinates in the semi-circle measured from $b\,c$, and through the points thus found draw the semi-elliptic line as shown.

Then, for the development No. 2, on the produced line through $s\,l$ in No. 1, set off from l to s', the distances that the points 1, 2, 3, etc., in the semi-ellipse $s4l$, No. 1, are from each other, and through the points thus found, draw lines parallel to the side $e\,l\,c$ of the cylinder. These lines will be the meridians shown on the front surface of the cylinder when it is laid out flat; for those on its near face, set off the same spaces from s' to l' as for the front face, and draw in the meridians as shown. For the development of the top and bottom edges, draw projectors through points 1, 2, 3, etc., in them, in No. 1, to cut the corresponding lines in No. 2; a continuous line drawn through the points of intersection—at both ends of the meridians—1, 2, 3, etc., will be the edges required. If the extreme meridians, $e'l'c'$, be put in full, as shown, the figure bounded by the lines $e\,d\,e'$, $c\,b\,c'$, and $e\,l\,c$, $e'l'c'$, will be the complete development of the curved surface of the oblique cylinder $d\,b\,c\,e$, No. 1.

If the student now compares this development with that of a portion of the surface of the right cylinder, lying between the two parallel but oblique section planes $a\,d$ and $e\,f$—the latter shown in dotted lines—in No. 1, Fig. 198, he will note that, though apparently similar, they are actually different in form and in outline, caused by one solid being perfectly *circular* in cross-section, while the other is elliptic.

91. From the development of the surface of the cylinder, and its frustums, we proceed to that of the cone. From the definition of this solid " as one generated by the revolution of a right-angled triangle about its perpendicular side as an axis," it will be understood at once, that, although all meridians drawn on its surface are, like those on a cylinder, actually of the same length, yet in the representation of them in elevation they are all, with the exception of the extreme bounding ones of the figure foreshortened, and of varying length, dependent upon their position. As errors are very likely to occur in any attempt to develop the surface of a cone, if this fact is not borne in mind, it is here referred to, to prevent a false development from being made by the student. As a first problem in connection with the cone and its development, we take the following—

Problem 97 (Fig. 200).—*Given the elevation of a right cone; to find the development of its curved surface.*

Now as the curved surface of a right cone may be conceived to be generated by the revolution of a straight line round an imaginary vertical axis, one end of that line moving in a circle described about the axis as a centre, while the other is in the axis itself, it follows that if a solid so generated be laid on a plane, and be caused to roll through

204   FIRST PRINCIPLES OF

one revolution, all the points in its base edge will together form the
arc of a circle having the apex of the cone as its centre, and the length
of its slant side as radius ; the *length* of the arc so struck or produced
being equal to the circumference of the base of the rolling cone. If,
then—as in the case of the cylinder—the cone be conceived to give off
its surface during one complete revolution, while rolling on a plane,
the surface of that plane actually rolled over by it, will be its developed surface.   To show this graphically will be to solve the problem
above given.

Let ABC, Fig. 200, be the elevation of the cone, A being its apex,
the line BC its base, and that through Aa its axis.   Then to find the
development of its curved surface, with a as centre, and half BC as
radius, describe a semi-circle, which will be a plan of the front half of
the cone.   Divide this semi-circle into any number of equal parts—say
eight—and number them 1, 2, 3, etc.   With A—the apex of the cone
—as centre, and AC its slant side as radius, describe from C—indefinitely—the arc of a circle, and on it from C, set off in C', *twice* the number of equal parts that the semi-circle on BC is divided into.   Join C'
with A, and the sector included by the two radii AC, AC', and the
arc CBC', will be the development required.

From the foregoing it will at once be seen, how the surface of a
frustum of a right cone may be readily found, for it is only necessary to set off vertically from the base line BC of the cone the intended
height or depth of the frustum, through the point thus given to draw
a line parallel to BC, as $b\,c$ in Fig. 200, and with A as a centre, and
Ac as radius, to describe an arc from c, cutting AC' in c', then the
figure included between the concentric arcs CC', $c\,c'$, and the lines C'c,
Cc, will be the developed surface of the frustum.

If the line of section of a right cone be *inclined* to its base, as $s\,l$ in
Fig. 200 is to BC, then the development of the frustum's curved
surface is found as follows : — Through the points of division
1, 2, 3, etc., of the semi-circle on BC, draw lines perpendicular to BC
to cut it in 1', 2', 3', etc. ; join these with A, the apex of the cone, by
right lines, and they will be meridians on its surface at these points.
Then, from the points set off on the arc drawn from C to C', draw
lines to A, and they will be the development of the meridians both on
the front and back surface of the cone.   Number these from C to B',
and B' to C' as shown.   On to the slant side, AC of the cone, project
over parallel to BC, the several points where the line $s\,l$ is cut by
meridians drawn from points 1', 2', 3', etc., in BC to A ; then from A
as centre, and the distances of the points just found in the slant side
AC from it as radii, cut the corresponding meridians drawn on the
development of the cone as shown, commencing on the one drawn
through AB', with As as radius, which will give the *lowest* point s'
in the curve.   The surface enclosed by a continuous line drawn
through the points thus found, with the arc CB'C', and the lines
Cl', Cl, will then be the development of the surface of the frustum
required.

92. As the oblique cone and its frustums play as important a part
in giving shape to flues, uptakes, funnel bases, etc., as the oblique

MECHANICAL AND ENGINEERING DRAWING   205

cylinder and pyramid, the solution of the following problem will show
how the development of their surfaces is found.

Problem 98 (Fig. 201).—*An oblique cone of a given diameter of
base and vertical height, has its axis inclined to its base at 45°;
required its elevation and the development of its curved surface,
together with that of a frustum of the same solid of a given height.*

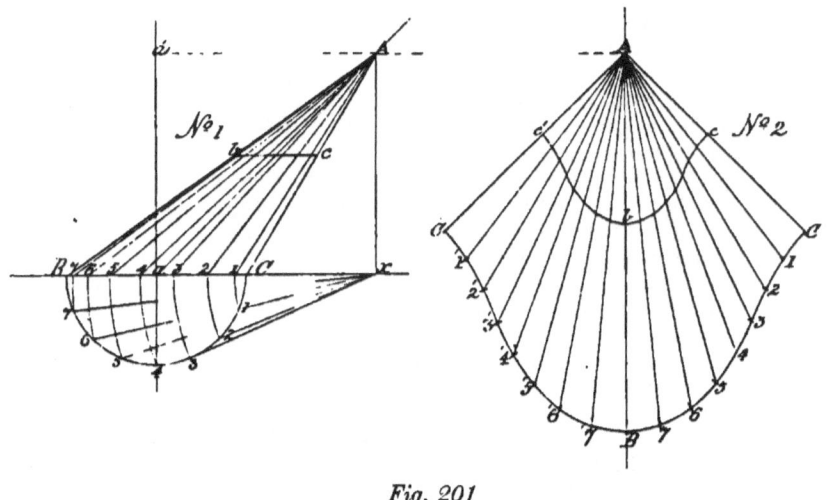

Fig. 201

Assuming the cone to be resting with its base on the HP, to draw
its elevation, take any convenient point, a in the IL as a centre, and
with half the intended diameter of its base as radius, describe a semi-
circle cutting the IL in B and C. At a draw a line, making with BC
an angle of 45°, and another perpendicular to it. On the perpendicular line set off from a in a' the intended vertical height of the cone,
and through a' parallel to BC draw a projector to cut the inclined line
from a in A, which will be the apex of the intended cone. Join
B and C with A by right lines, and the triangle ABC will be the
elevation.
To find the development of its curved surface, divide the semi-
circle drawn on BC into any number—say eight—of equal parts, and
from A let fall a vertical projector to the IL—or base BC of the cone
produced—to cut it in x. Now it is evident that the longest and
shortest meridians that can be drawn on the surface of the cone, will
be its bounding lines AB and AC; then to determine the lengths of
any intermediate ones—for that is what is required to be known—
proceed as follows :—
From x in BC produced, draw right lines to the several points of
division in the semi-circle, and take each of these lines as a radius,

220   FIRST PRINCIPLES OF

and x as centre, and describe arcs to cut BC in points 1, 2, 4, etc.
Join each of these by right lines with A, the apex, then will they be
—taken in order—the actual lengths of meridians drawn on the front
surface of the cone, from A its apex to the points 1, 2, 3, etc., in its
base edge. The axis of the cone being in a plane parallel to the VP,
or plane of the paper, the meridians on its back surface will be of the
same length as those on the front one, being directly behind them.

93. To make the finding of the actual lengths of meridians on an
oblique cone's surface as clear as possible to the student, let him conceive a right-angled triangular plane (having a *constant* altitude, or
perpendicular, as $A\,x$ in the figure, but a *varying* base, of lengths equal
to the distances between x and the points 1, 2, 3, etc., in the semi-
circle on BC) to swing on its perpendicular edge $A\,x$ as a hinge,
through the arcs drawn between the points 1, 2, 3, etc., in BC, to the
corresponding one in the semi-circle drawn on BC, it will then be
seen that the hypothenuse of such a trianglar plane would coincide
with, and be of exactly the same length, as a line drawn on the cone's
surface from its apex to the point in its base corresponding with that
of the particular plane taken. For instance, take the plane having a
base $x3'$ in No. 1, then the line A3′ will· be the actual length of
a meridian drawn on the surface of the cone from point 3 in its base
edge to its apex, and so with any other of the planes.

Having thus found the length of any meridian drawn on the cone's
surface, its development offers no difficulty. To find it, draw a line indefinitely in No. 2, Fig. 201, parallel to $A\,x$ in No. 1, and project over
to it the point A ; on this line set off from A the distance AB, equal
to the length of AB, No. 1. On either side of B, No. 2, draw an arc of
a radius equal to the length between B and point 7 in the semi-circle
on BC in No. 1, and from A in No. 2, with the distance between A
and point 7 in BC, No. 1, as radius, draw arcs cutting those struck
from B in points 7, 7′; and join these points with A by faint lines as
shown. From 7 and 7′ as centres, strike arcs of the same radius as
from B ; and from A with A6′ in No. 1 as radius, cut the arcs last
drawn in points 6, 6′; join these points with A by right lines, and repeat
the process till AC in No. 1, as radius, is reached ; then join C and C′
with A, and the figure AC′BCA will be the complete development of
the whole surface of the oblique cone, of which ABC, No. 1, is the
elevation.

For the development of the surface of the frustum bBCc, No. 1, of
the same cone, we have in effect to cut off from that just found so
much of the surface as will cover the smaller oblique cone in No. 1,
having the line of section $b\,c$ for base, and A for vertex. With the
meridians already drawn in, all that is required to be done is to
measure off on each of them respectively, its length from the vertex A
to· the point in $b\,c$ in No. 1, where that meridian crosses it, and
transfer it to its corresponding meridian in No. 2 ; then a continuous
line, as $c'b\,c$, drawn through the points thus found, will be the development of the top edge of the frustum, giving the remaining lower
part of the figure in No. 2 as the required development of the surface
of the frustum bBCc, given in elevation in No. 1, Fig. 201.

MECHANICAL AND ENGINEERING DRAWING   207

94. Having shown how to find the development of the surfaces of
the cylinder and cone—right or oblique—with that of their frustums,
we have now to explain how the covering of a "sphere" is obtained.
Its surface, *as a whole*, has been previously stated to be non-
developable, but it does not necessarily follow that such a solid cannot
be covered. The impossibility of doing this with *one* sheet of material
is overcome in practice by dividing its surface into what are called
" gores," or figures which may be defined as made up of two spherical
triangles joined at their bases ; such a triangle differing from a plane
one in having all its sides curved—they being arcs of *great circles* of
the sphere—instead of being straight.
In previous problems in connection with the sphere, it was shown
that any section of it made by a plane passing though its axis, gave
a great circle—or one equal in diameter to the sphere—for that section;
and from the fact, that such a solid is generated by the revolution
of a semi-circle about its diameter as an axis, it follows that any plane
section of it, at right angles to a great circle section, will be a circle
having a diameter proportioned to the distance that the cutting plane
producing it, is from the centre of the sphere. Then as the halves of
a sphere—or hemispheres—are equal solids, the covering of only one
half need be found.
The parts into which a sphere is supposed to be divided for the
purpose of finding its covering, are such as would be produced by a
series of vertical planes passed through its axis, dividing its equatorial
circumference into equal parts, and giving what are called "lunes" as
the resultant solids ; and it is the development of the surfaces of these
lunes which has to be found. The first problem in this connection is—

Problem 99 (Fig. 202).—*Given the plan and elevation of a hemisphere ;
to find its approximate covering, the axis, or pole, being assumed
to be vertical.*

Let Nos. 1 and 2, Fig. 202, be the given plan and elevation of the
solid, resting with its base AB, on the HP, and its pole Pp, perpendicular to it. First divide its surface in No. 1 into, say six equal parts,
by meridians AB; 1 1'; 2 2'; passing through the pole P, and find
their elevations as in No. 2. Now the meridian lines passing through
P are the plans of meridian planes cutting the hemisphere into
six half-lunes, and as these are equal solids, having similar and equal
surfaces, it is only necessary to find the covering of one of them.
To do this, divide the quadrantal arc AP in No. 2, into, say
four equal parts, in points 3, 4, 5, etc., and through them draw
lines parallel to AB, to meet the opposite arc PB. Assume these lines
to be parallels drawn on the surface of the hemisphere, and find their
plans—which are circles—in No. 1. Number the points where they cut
the meridians 2 2', 1 1', to correspond with their elevations. From the
point x, in the base of the hemisphere in No. 1, on the axial line Pp
produced, set off a length $x\,p$, equal to the *actual* length of the arc AP
in No. 2; divide this into the same equal parts that AP is divided
into, and through the points of division, draw lines at right angles

208.   FIRST PRINCIPLES OF

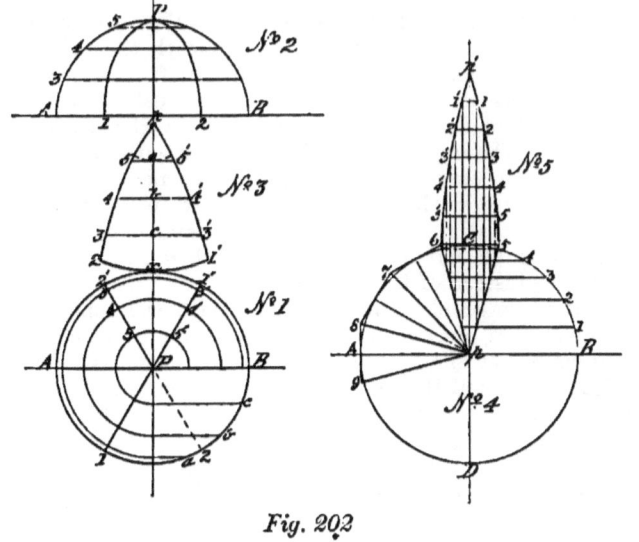

Fig. 202

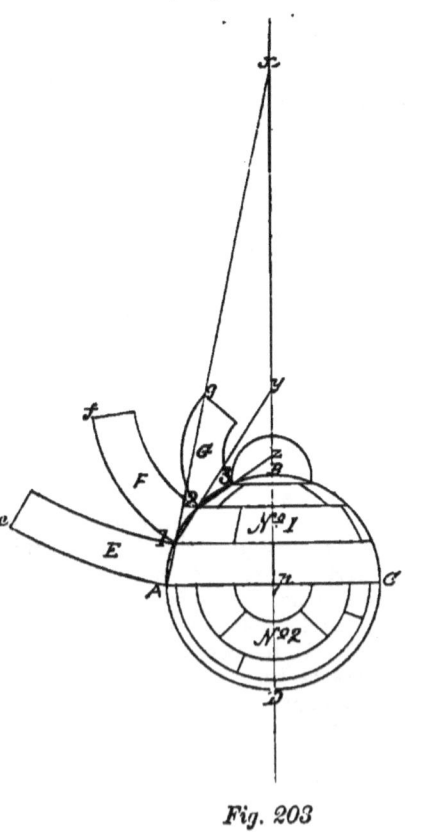

Fig. 203

MECHANICAL AND ENGINEERING DRAWING   209

to Pp. On either side of Pp, on these lines, set off in points 3 3', 4 4', 5 5', the actual lengths of the arcs drawn between the meridians 2 2', 1 1', in No. 1 ; and through them and p, draw curved lines. Make the length of these—measured along them from p to 1', 2—equal to p x, and through 2, x, 1', draw the curved line shown. Then the spherical triangle p2 1', will be the development of the surface of one-sixth of the hemisphere.

The greater the number of gores, or parts the surface of the hemisphere is divided into the nearer will they—when properly bent and joined together—approach the true spheric form.

95. In practice, where a hemispherical surface is required, as, for instance, in egg-ended boilers, tops of floating buoys, dome-shaped coverings, etc., it would not be possible to make it of plates cut to the shape of a *perfect* gore, as the plates could not be riveted together at their upper or pointed ends. In such case, it is usual to cut the gores short of the required length, and fit a dished circular crown plate to which the upper ends of all the gores are riveted. In the hemisphere given in the problem this crown plate would about equal in diameter the smallest circle shown in the plan No. 1, necessitating the gores being cut at their top ends, to the curve of an arc—shown dotted—struck from p, of a radius to suit the crown plate.

In many curved surfaced structures of hemispherical form—to save expense and much labour in construction—the covering adopted is that of a spherical polyhedron, having a great number of faces ; each face being tangent to a hemispherical surface, the touching part being a straight line, coinciding with, and falling on a meridian, which divides the face into two equal parts throughout its length. In the case of the covering of a truly hemispherical surface, each gore would not only have to be bent lengthwise, to the arc of a circle of the radius of the sphere, but it would also require to be bent crosswise, or dished, which would necessitate special care, so as to ensure the same curvature in all. By adopting the polyhedron surface, the covering material only needs bending in one direction—lengthwise—as each gore is a part of a cylindric surface and therefore developable. To develop such a surface proceed as follows—

Let the circle ABCD, No. 4 (Fig. 202), be the plan of a hemisphere to be covered by a surface of the form of a spheric polyhedron of twelve faces or sides, and let the lines AB and CD—which are at right angles to each other—be meridians on the hemisphere. Divide the quadrant AC into six equal parts, and at the points of division draw faint lines to the centre, or pole p. Also divide the arc CB into the same number of equal parts, numbering the points of division 1, 2, 3, etc., as shown in No. 4. From p, draw radials through points 5, 6 ; and at C, a tangent to the circle, to cut them ; then will the triangle p, 5, 6 in No. 4, be the plan of one of the gores of the given surface.

To find its development, parallel to the diameter AB No. 4, and through the points 1, 2, 3, etc., in the arc CB, draw lines to cut the two radials p 5, p 6 ; produce the diameter CD indefinitely—upwards —and on it from C, set off to p', the equal parts into which the arc CB

P

210   FIRST PRINCIPLES OF

is divided, and through the points of division draw lines parallel to the
tangent drawn through C. Then from the points of division in the
radials $p\,5$, $p\,6$, cut by the lines drawn from points 1, 2, 3, etc., in the
arc CB No. 4, draw projectors parallel to CD produced, to cut the lines
drawn through it in the points 1 1', 2 2', 3 3', etc. Through these last-
found points and p', draw curved lines as shown, and the triangular
figure $p'6\ 5$ will be the development of the gore required.

96. If a hemispherical or dome-shaped surface is of large area,
a different method of covering it would be resorted to. Instead of
dividing the spherical surfaces into gores, that of "zones" or circular
belts cut into convenient-sized sections would be the form given to the
material. The method of finding the proper shape of such sections is
shown in Fig. 203, and is as follows—

With p, in the given straight line AC in the figure as centre, and
with pA as radius, describe the circle ABCD. Let the upper half of
it, No. 1, be the elevation, and the lower half, No. 2, the plan, of the
front part of a given hemisphere, whose surface is required to be
covered by material in the form of zones or belts. Divide the arc AB,
of the semi-circle ABC, into any number of equal parts, say four; and
at the points of the division 1, 2, 3, draw lines parallel to AC.
Produce the diametral line BD—upwards—indefinitely, and through
the points A, 1, in the arc AB, draw a line to cut this produced line in
x. Through points 1 and 2, and 2 and 3, draw similar lines to cut
BD produced in y and z. Then with xA, and x1, as radii, draw arcs
indefinitely, and repeat the process with y and z as centres, and y1, y2 ;
z2, z3, as radii. Now the surface enclosed between each of the pairs
of concentric arcs, struck from x, y, z, as centres, are portions of the
surfaces of frustums of cones, having those centres as their apices, and
the lines AC, 1 1', 2 2', 3 3', as their bases; the heights of the frustums,
and the slope of their sides, being determined by the number of parts
into which the arc AB of the semi-circle ABC is divided.

The lengths of each of the covering pieces in the zones or belts are
determined by the number into which the belt is to be divided. Let
this number be four; then to show their position and length on the
spheric surface, find by projection on the plan of the front half of the
hemisphere in No. 2, the plans of the lines 1 1', 2 2', 3 3' in No. 1, which
will be semi-circles as shown. Now the radial line pD divides each of
the semi-circular belts into two parts, and as four such parts of each
belt will cover the hemisphere—with the exception of the crown plate
—the length of one of them in each, taken in order, will be the length
that each of the strips E, F, G, in No. 2, must be. Therefore on each
of the *outer* arcs drawn from x, y, z, set off the actual lengths in
points, d, f, g, of their corresponding arcs in the quadrant Ap, D,
No. 2; and through d, f, g, draw lines radiating to x, y, z, to cut the
inner arcs, or edges of each belt as shown.

In arranging the plates for covering such a surface, they would for
strengthening and other important reasons be made in practice to
"break joint," as shown by the short radiating thick lines in the half-
plan of the hemisphere No. 2 in the figure.

By one or the other of the methods shown and explained in

MECHANICAL AND ENGINEERING DRAWING   211

Figs. 202-3 the coverings of the sphere—oblate or plolate—the ellipsoid, paraboloid, and hyperboloid with their frustums may be found. Combinations of parts of these solids, with those of the cylinder, cone, and sphere, constantly occur in practice, and all that is required of the student for the mastery of each case as it arises, is to make himself thoroughly acquainted with the actual forms of the solids which enter into combination, and then apply the principles which have been so fully explained in this and previous chapters, and on which their correct delineation depends. .

97. With the problems on the covering of the sphere, etc., the subject of the " Development of the Surfaces of Solids " is concluded, and with it the exposition of the principles of that special kind of "projection" on which the art of Mechanical and Engineering Drawing is based. The problems in each division of the subject might have been considerably increased in number, but as such an extension would have involved the expenditure of more of the student's time than the subject warrants, as many have been given as will be found necessary for all his future requirements.

A careful study of the foregoing first principles of the Mechanical Draughtsman's art, and the conscientious working out of all the problems furnished for their complete elucidation, will lead the student to an easy comprehension of the method of their practical application to the delineation of all kinds of machine elements, and engine details, which may form the subject of a further work by the author of the present one here concluded.

www.ingramcontent.com/pod-product-compliance
Lightning Source LLC
Chambersburg PA
CBHW031813230426
43669CB00009B/1120